A "Hands On" approach to teaching...

Logic

Linda Sue Brisby
Andy Heidemann
Natalie Hernandez
Jeanette Lenger
Ron Long
Petti Pfau
Scott Purdy
Sharon Rodgers

HANDS ON, INC.
SOLVANG, CALIFORNIA

Layout and Graphics: Scott Purdy
Primary Layout: Linda Sue Brisby
Illustrators: Petti Pfau, Sharon Rodgers, Scott Purdy, and Linda Sue Brisby
Cover Art: Petti Pfau

Printed in the United States of America.

Order Number: HO 103

HANDS ON, INC.
2121 Rebild Drive
Solvang, CA 93463

Table of Contents

Logic
Task Analysis

Attributes	1. Sorts and classifies objects by color.	Primary
Attributes	2. Sorts and classifies by size.	Primary
Attributes	3. Sorts and classifies by shape.	Primary
Attributes	4. Sorts and classifies by thickness.	Primary
Attributes	5. Sorts and classifies objects by use.	Primary
Attributes	6. Sorts and classifies objects by component parts.	Primary
Attributes	7. Sorts and classifies by two or more attributes.	Primary
Sequencing	8. Identifies the next item/step or series.	Primary
Sequencing	9. Puts items into a logical sequence.	Primary
All/Some/None	10. Demonstrates the concept of all/some/none.	Primary
And/Or/Not	11. Demonstrates the concept of and/or/not.	Primary
If/Then	12. Demonstrates the concept of if/then.	Primary
Relationships	13. Identifies relationships among objects.	Primary
Attributes	14. Identifies relationships between two objects according to relative attributes.	Middle
Sequencing	15. Identifies simple relationships among objects according to position in a pattern or sequence.	Middle
Venn Diagrams	16. Uses the techniques of Venn Diagrams to represent relationships.	Middle
All/Some/None	17. Demonstrates an understanding in using the words ALL, SOME, NONE.	Middle
Valid/Invalid	18. Identifies valid or invalid information and/or conclusions.	Middle

Relationships	19. Develops simple relationships among objects and draws conclusions based upon certain qualities or conditions.	Middle/Upper
Elimination	20. Solves a problem based on disqualification of available options.	Middle/Upper
And/Or/Not	21. Demonstrates an understanding in using the words AND, OR, NOT	Middle/Upper
Interpolates	22. Interpolates information into a set of data.	Middle/Upper
Extrapolates	23. Extrapolates information from a set of data.	Middle/Upper
Sequencing	24. Identifies relationships among objects according to position in a pattern or sequence.	Middle/Upper
Syllogisms	25. Explains logical relationships by writing accurate syllogisms.	Middle/Upper
Venn Diagrams	26. Uses Venn Diagrams to represent complex relationships.	Upper
Attributes	27. Identifies relationships among three or more objects according to three or more attributes.	Upper
Sequencing	28. Identifies complex relationships among objects according to position in a pattern or sequence.	Upper
Relationships	29. Develops relationships among objects and draws conclusions based upon certain qualities or conditions.	Upper
Elimination	30. Solves complex problems based on disqualification of available options.	Upper
If/Then	31. Demonstrates an understanding in using the words IF...THEN, IF...NOT, IF AND ONLY IF...	Upper

Introduction

This is the third book in a series which Hands On, Inc. is writing to help teachers more effectively teach math by using "manipulative" or "hands on" materials. We feel that children learn most effectively when they have concrete experiences to solidify the abstractions of mathematics.

Each of our books focuses on a math "strand" identified by the State of California Mathematics Framework. Already completed are books on Statistics, Probability, and Graphing and on Measurement. Future books will focus on Pattern and Function, Geometry, Algebra, Number Theory, and Problem Solving.

In preparing this book we were amazed at the lack of articulated and organized information on logic in the elementary grades. Many books provide logic puzzles and games, but we feel that this publication is unique in this field. We have provided a complete task analysis of logic skills for kindergarten through ninth grade students, which breaks the concept (subject) down into its component parts -- a logical exercise in itself. We researched the area of logic thoroughly, all the time trying to break down the "jargon" into understandable and useable terms.

We found some very fine lines separating Logic from Pattern and Function. We have identified three essential understandings which differentiate these two areas. This is not to say that these strands need to be, or should be, taught separately. On the contrary, we feel strongly that integration of the strands is essential. We do feel, however, that the teacher must have a clear understanding of the components of logic in order to teach it effectively. We make this distinction as follows:

LOGIC:

USES ORGANIZED DATA TO DRAW NEW CONCLUSIONS

1. Classifies and sorts by attribute.
2. Evaluates statements in order to make conjectures and to draw conclusions.
3. Creates a chain of logical arguments to reach a valid conclusion.

PATTERNS AND FUNCTION

PLACES ORDER IN DISORDER

1. Creates a rule that enables the pattern to extend indefinitely.
2. Extends patterns and relationships to find quantities and relationships (e.g. - tables).
3. Creates a variety of ways to view the same pattern.

As you work through the lessons presented in this book, we suggest that you review these differentiations. When you work with students, you will find that the two strands will begin to segregate themselves.

You will also find that this book offers many lessons which are seemingly out of the area of math. Certain task analysis items such as the use of AND, OR, NOT and IF...THEN lend themselves as much to language arts as to numbers. We see this as an added bonus in terms of curriculum integration. One of the concerns we hear from teachers is, "How can I fit all of this into my teaching day?" Curriculum integration provides an answer. The field of logic is not confined to the area of math. Its application is universal and its study is necessarily wide. Many of the logic lessons presented in this book will allow you to combine a math skill with a social studies, science, art, or language arts skill. We encourage this approach.

In terms of the task analysis we created for logic, we found that it was almost impossible to create the type of hierarchy of skills that we presented in our first two books. Logical thought is not necessarily a "step by step" procedure in which one skill is mastered before moving on to the next level. Based upon this, we encourage you to look beyond the primary, middle, and upper labels which we have used. With minor alterations, most of the lessons in this book can be used at almost any grade level, including high school. We hope you will read all of the lessons and enjoy using them with your students.

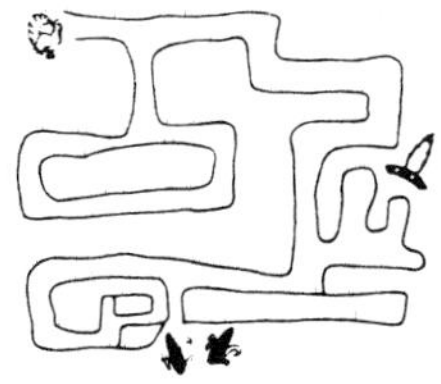

The Logic of Mazes

Mazes are a most intriguing, fun way for children to develop their logic skills.

Teachers and children alike are often unaware of the process used in determining the outcome in these activities.

The mazes we have created need to be used in ways that draw attention to these processes and they must be explained with the terminology of logic. Using these mazes, kindergarten, first, and second grade teachers have discovered that they help students grasp a logical, "common sense" approach to problem solving.

We have increased the difficulty of the mazes as the children become more sophisticated in logic.

There is one maze provided after each month of lessons. The theme of the maze is coordinated with the major theme of that month. The mazes can be used at anytime during each month.

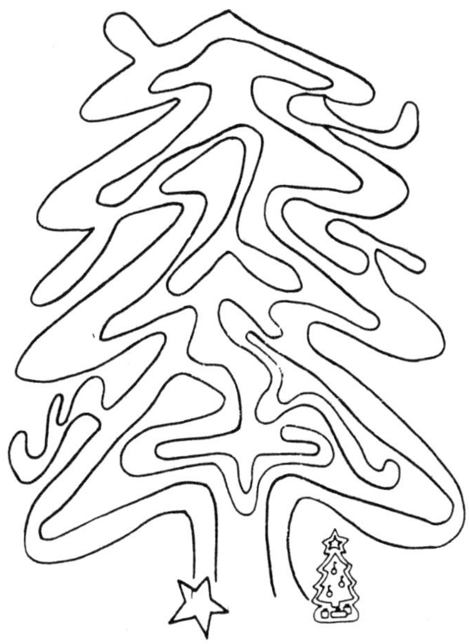

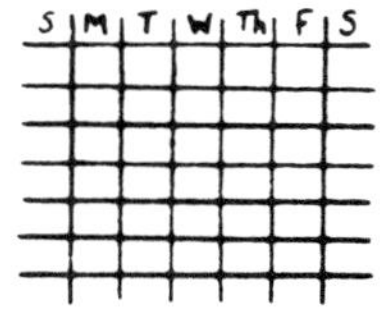

Calendar Logic

September

Grade Level: Primary

TASK ANALYSIS: 6 – Sort and classify objects by components
8 – Identifies next item/step in a series
11 – Demonstrate the concept of or/and/not
12 – Demonstrates the concept of if/then

MATERIALS: Parts of calendars: dates, days, months, years – grid for calendar, glue, monthly calendar on wall

ORGANIZATION: Whole class activity for opening
K – 3: 20 – 30 minutes in partners
Use in conjuction with Hands On Measurement calendar activities

PROCEDURE:
- Have the children work in partners to sort and classify the calendar parts.
- Make a calendar.
- Each day when doing the calendar have a child (children) tell the next date on the calendar and explain: "Today is Tuesday, January 10, 1989." "Yesterday was Monday, January 9, 1989." "Tomorrow will be Wednesday, January 11, 1989."
- Ask questions: "Can we put up the 23rd or the 24th?" "Why?"
- "Is the year 1988 and 1989?" "Why or why not?"

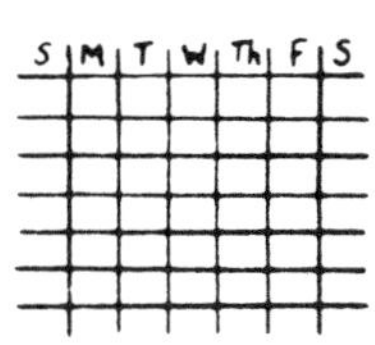

Apple Logic 1

September

Grade Level: Primary

TASK ANALYSIS: 1 – Sort and classify objects by color
2 – Sort and classify objects by size
7 – Sort and classify objects by two or more attributes
10 – Develops concept of all/some/none

MATERIALS: Apples (1 per student), butcher paper graph or floor graph, paper tent headings for labelling the categories, marker

ORGANIZATION: Whole class activity for 1 to 2 days
K – 1: 20 to 30 minutes
2 – 3: 30 to 40 minutes

PROCEDURE:
- Make paper tent headings for the colors of apples.
- Place tents on the floor graph.
- Have students "get-to-know" apple by discussing its attributes (color, size, bumps, bruises).
- Ask students to place apples on the graph in the color catagories.
- Discuss the data and ask students to use sentences with all/some/none.
- Change the tent headings and ask the children to catagorize according to size.
- Discuss the new data using all/some/none.
- Ask the children to sort and categorize the apples using two or more attributes.
- Discuss the data, (i.e. "All the large apples are red," "All the small apples are yellow," "Some of the green apples were small," "None of the red apples were small," etc.).

Apple Logic 2

September

Grade Level: Primary

TASK ANALYSIS: 8 – Identifies next item/step in a series
9 – Put items into logical sequence

MATERIALS: Apples (either supplied by students or teacher), cooking utensils

ORGANIZATION: Whole class or smaller cooperative groups
K – 3: 20 – 30 minutes (several days)
Easily integrated with Science and Social Studies.

PROCEDURE:
- Wash apples.
- Cut apples into quarters.
- Fill crock pot with water 1/8 cup.
- Pour apples into crock pot.
- Cook overnight.
- Next day mash through colender.
- Sprinkle cinnamon on each serving.
- Be certain all children are active in each step of the procedure. Discuss what sequence or steps were followed.
- Discuss why a specific sequence was necessary to obtain the result of smooth applesauce.

Hand Game Logic

September

Grade Level: Primary

TASK ANALYSIS: 8 – Identifies next item/step in a series
12 – Demonstrates the concept of if/then

MATERIALS: At least 3 of the same item in two colors (marbles, pattern blocks, beans, etc.)

ORGANIZATION: K – 3: Partners
20 – 30 minutes

PROCEDURE:
- Show items to the children: example – 2 red marbles and 1 blue marble.
- Shake the items in your hands.
- Divide items between both hands.
- Show what is in one hand, but keep other hand closed.
- Ask, "If you see the blue one in this hand, then what is in the other hand?"
- "How do you know?"
- Repeat several times, using all combinations.
- Have children work in pairs.

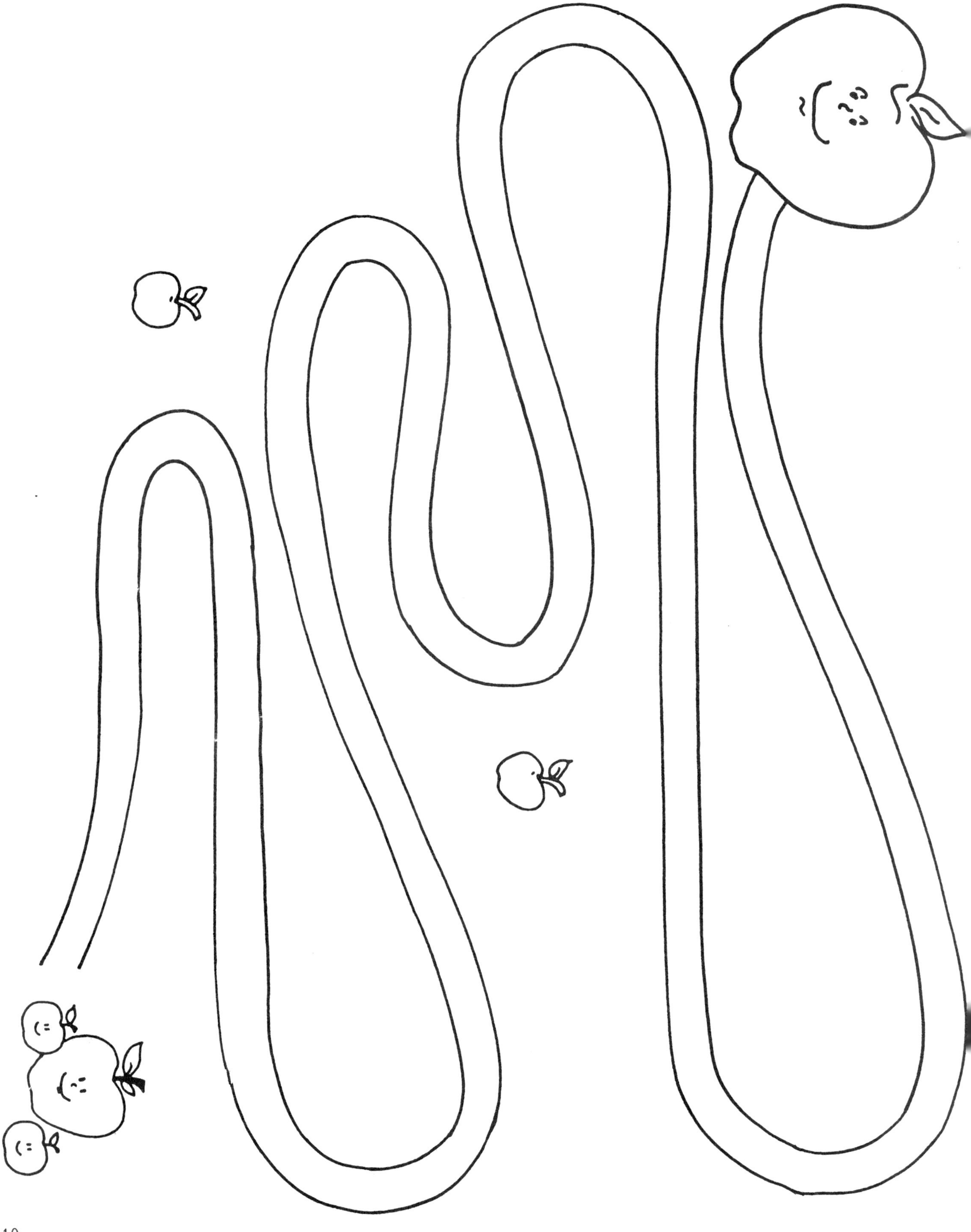

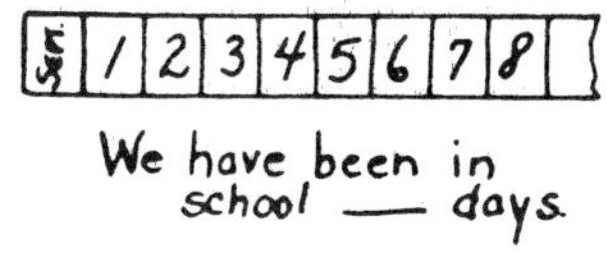

Number Line Logic

October

Grade Level: Primary

TASK ANALYSIS: 8 – Identifies next item/step in a series
12 – Demonstrates the concept of if/then

MATERIALS: Number line (see Hands On Measurement) marker, days in school graph

ORGANIZATION: Whole class activity
K – 3: Every day during opening activities

PROCEDURE:
- Count the days in school
- Ask the children what number comes next.
- Write down all the children's ideas on the chalkboard.
- Go over each number given and ask for an explanation whether it should go on the number line.
- Include dialog such as: "If all the 20's start with a 2, then 27 should be written 2 7."

Number line

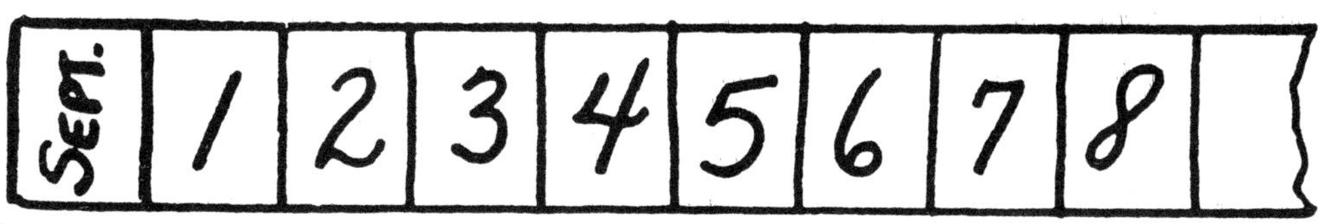

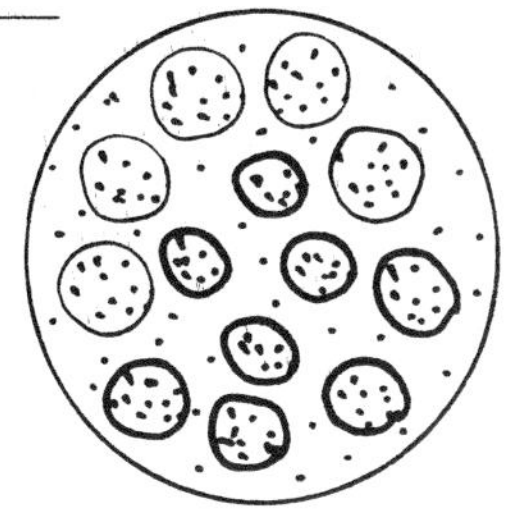

Make and Bake Mini Pizzas

October

Grade Level: Primary

TASK ANALYSIS: 8 – Identifies next item/step in a series

MATERIALS: English muffins, pizza sauce, cheese, (any other ingredients to be put on pizza)

ORGANIZATION: Whole class activity
K – 3: 20–30 minutes for lesson
Another 20 to 30 minutes to make and bake mini pizzas

PROCEDURE:
- Brainstorm for items needed to make mini-pizzas.
- Have items ready, and put in middle of circle.
- "What do we need first? Why?"
- "What comes next? Why?"
- Continue until all items are used on the model mini-pizza.
- Children then make their own mini-pizzas according to their discussion; bake and enjoy.

EXTENSION: A great activity to integrate social studies and science with a unit about Columbus.

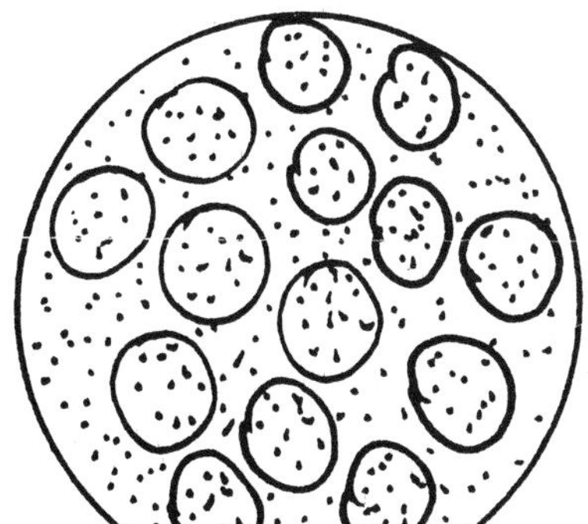

Carve the Class Pumpkin

October

Grade Level: Primary

TASK ANALYSIS: 2 – Sort and classify objects by size
3 – Sort and classify by shape
7 – Sort and classify objects by 2 or more attributes

MATERIALS: Pumpkins of various shapes and sizes, knife

ORGANIZATION: Whole class activity
K–3: 20–30 minutes

PROCEDURE:
- If possible,take a field trip to a pumpkin patch; each child may purchase a pumpkin, or have each child bring a pumpkin from home.
- Sort and classify the pumpkins by size.
- Sort and classify the pumpkins by shape.
- Sort and classify the pumpkins by 2 or more attributes.
- At the pumpkin patch choose a class pumpkin.
- Carve the class pumpkin.
- Have the children discuss what happens first, next and why.
- Continue carving.

A "Ghostly" Pattern

October

Grade Level: Primary

TASK ANALYSIS: 10 – Demonstrate the concept of all/some/none
11 – Demonstrate the concept of or/and/not
12 – Demonstrate the concept of if/then

MATERIALS: 1 work sheet per child, pattern blocks, pencil

ORGANIZATION: K – 3: station activity
10 – 15 minutes

PROCEDURE:
- With the entire group, demonstrate how to do the worksheet.
- Children will do paper during independent time.
- After all the children have finished the paper, have some children give a sentence using all/some/none.
- Ask questions, such as, "Did anyone use all of one kind of pattern block?" "Why?" "Who used some (squares, hexagons, etc.)?" "Why?"
- Have some children give a sentence using or/and/not.
- Ask questions such as "Did anyone use (hexagons) or (trapezoids)?" "Why?" "Did anyone use (hexagons) and (trapezoids) and not (triangles)?" "Why?"
- "If we used all (hexagons), then would the figure be covered?" "Why?"
- "What should we use to cover the entire figure?"

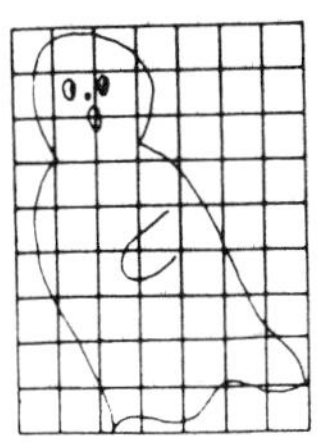

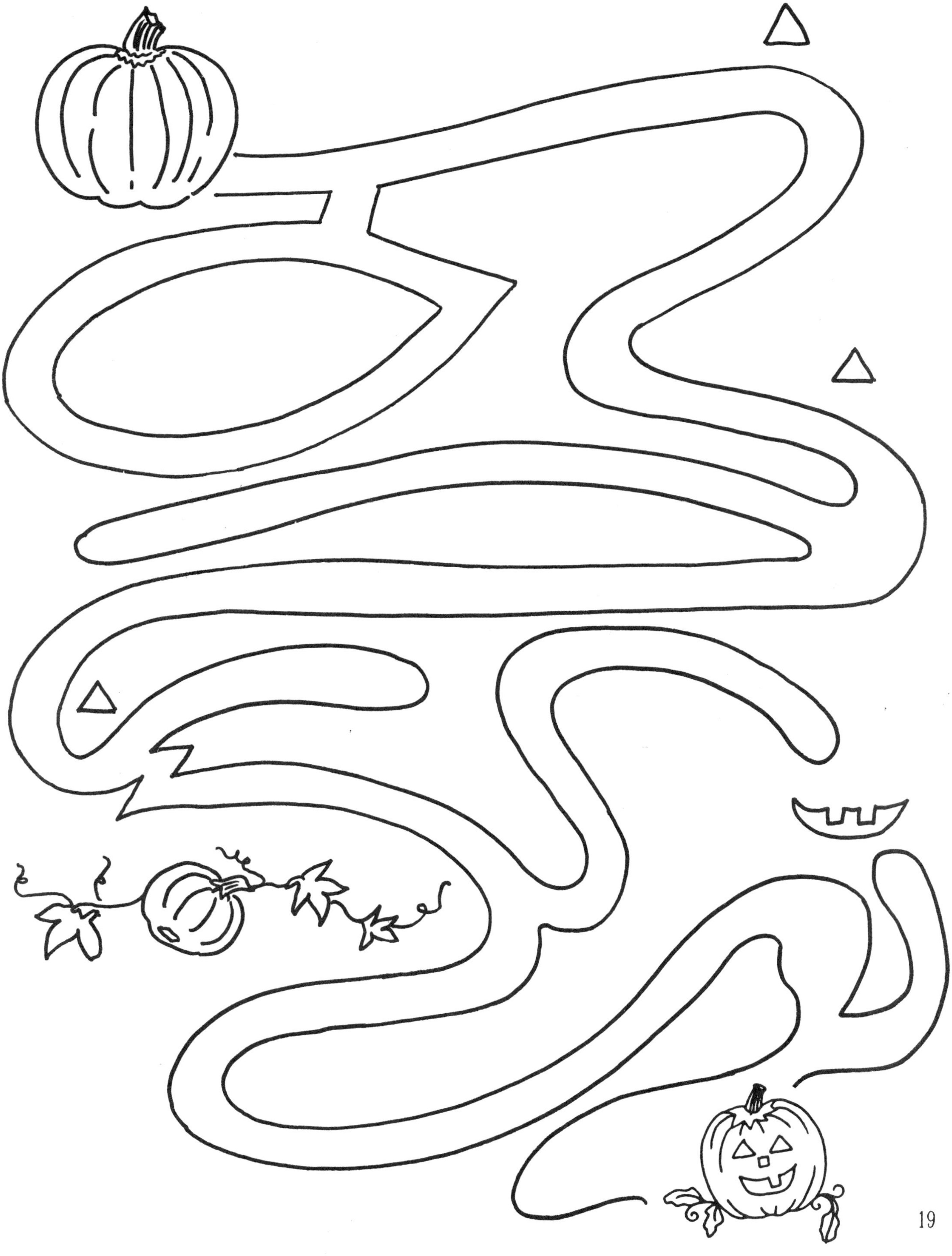

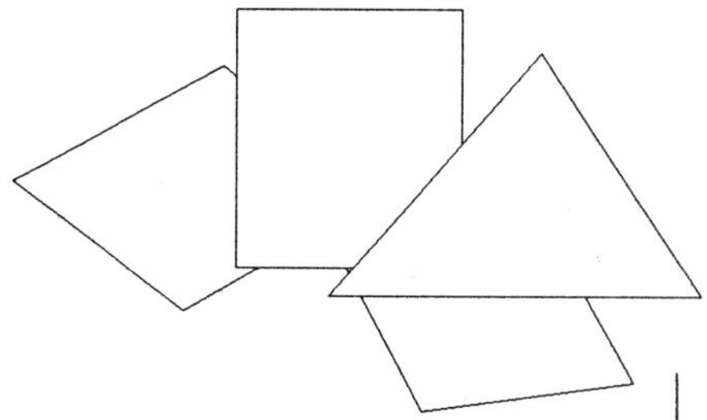

Pattern Block Logic

November

Grade Level: Primary

TASK ANALYSIS: 6 – Sort and classify by components
10 – Demonstrate the concept of all/some/none

MATERIALS: Xerox copy of desk top graph with headings (1 per student), pattern blocks (red, green, blue), floor graph 24" x 36

ORGANIZATION: Whole class activity (later in year small cooperative groups)
K – 1: 20 to 30 minutes
2 – 3: 40 to 50 minutes

PROCEDURE:
- Distribute pattern blocks to each student.
- Have students place blocks on floor graph by attributes (color, number of sides, shape).
- Discuss classifications:
  "Some blocks have four sides."
  "All the green blocks have three sides."
  "None of the green blocks have four sides."
- Follow same procedure on individual desk-top graphs.
- Have students explain rules used to sort and classify. Discuss classifications.

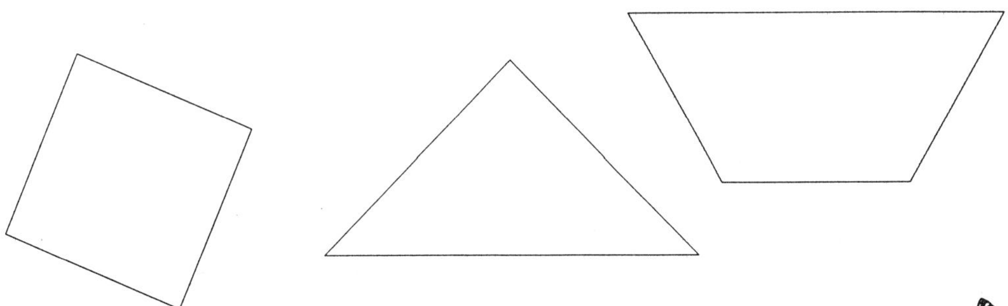

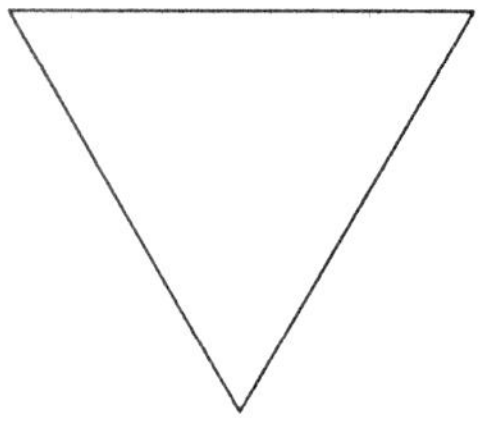

Pattern Block Logic

Corn Logic

November

Grade Level: Primary

TASK ANALYSIS:
1 – Sort and classify objects by color
2 – Sort and classify objects by size
3 – Sort and classify objects by shape
4 – Sort and classify objects by thickness
7 – Sort and classify objects by 2 or more attributes
9 – Put items in logical sequence (small to large)
10 – Demonstrate the concept of all/some/none

MATERIALS: Indian corn, sweet corn, popcorn, mini-pickled corn, frozen corn, canned corn, butcher paper, marking pens

ORGANIZATION: Whole group activity
K – 3: 20 – 30 minutes

PROCEDURE:
- Make butcher paper graph with headings.
- Discuss each heading.
- Children graph according to attributes.
- Discuss and interpret data.
- Place items in logical sequence (largest to smallest, smallest to largest, etc.).
- Ask questions all/some/none.

		 brave	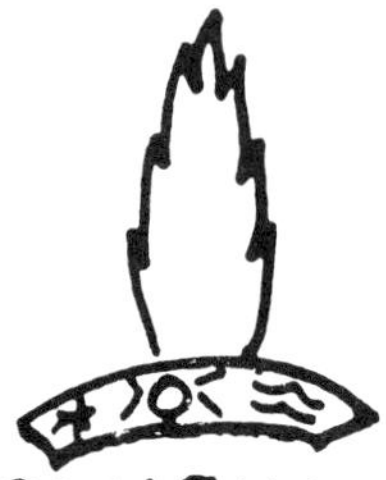 squaw

Headdresses

November
Grade Level: Primary

TASK ANALYSIS: 3 – Sort and classify objects by shape
6 – Sort and classify objects by components

MATERIALS: Construction paper bands for Indian girl and boy, feather pattern, Pilgrim boy and girl hat, pattern, stapler, yarn, scissors, crayons, glue, junk boxes of buttons and beads, butcher paper, marking pens, headings

ORGANIZATION: Whole class activity
K – 3 for presentation: 20 – 30 minutes
Cooperative groups for constructing hats: 30–40 minutes
Discussion (whole class activity) to graph: 20–60 minutes

PROCEDURE:
- Introduce children to the four different head-dresses.
- Children make their choice.
- Discuss shape and component parts of each head-dress.
- Using data, children can graph by shape, and/or by component parts.
- Students may graph the number of children that chose each head-dress.
- Using component parts, children may demonstrate all/some/none.
- Ask questions such as: "Do all/some/none...tie? slip on head? have feathers? cover whole head? keep head warm? have buttons? have beads? should be asked.

Pilgrim Hats

Diameter: 11"
Brim: 2 1/4"
Crown: 3 3/4"
Buckle: 3 1/2" square
Cut out on all the lines

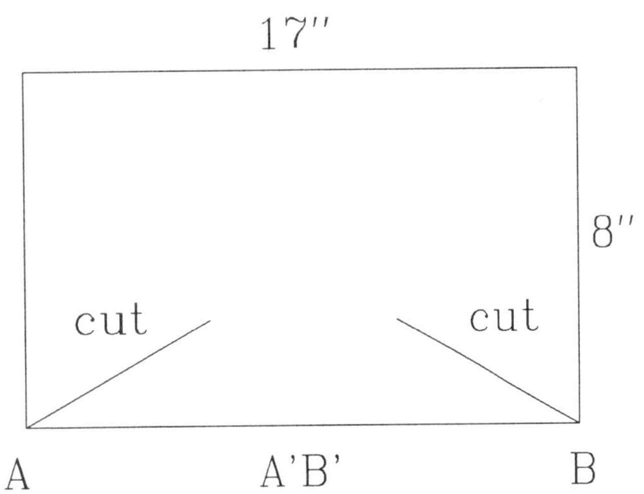

Cut out on lines.
Fold A to A' and staple
Fold B to B' and staple
Put yarn on ends of hat and tie under chin.

What Do You Like Best in School?

November

Grade Level: Primary

TASK ANALYSIS: 10 – Demonstrate the concept of all/some/none

MATERIALS: A butcher paper graph, felt tip pen, several subject headings, colored dots

ORGANIZATION: Whole class activity
K – 2: 20 – 30 minutes

PROCEDURE:
- Have the graph prepared.
- Place on board.
- Discuss the subjects on the graph with the children.
- Have the children put their names or colored dots under their two favorite subjects.
- This may be a whole group activity or done independently by children throughout the day.
- After all data has been graphed, gather as a whole group.
- Discuss the findings: "All the children like reading," "Some like math," and "None like recess."

Buttons

December

Grade Level: Primary

TASK ANALYSIS:
1 – Sort and classify objects by color
2 – Sort and classify objects by size
3 – Sort and classify objects by shape
7 – Sort and classify objects by two or more attributes
9 – Put items into logical sequence

MATERIALS: Buttons and sorting mats

ORGANIZATION: K–1: 20 to 30 minutes; work in pairs
2–3: Cooperative groups

PROCEDURE:
- Give each pair a handful of buttons and a sorting mat.
- Ask students to sort as follows:
  1. by color
  2. by size
  3. by shape
  4. by any two attributes, each time telling how they are sorted.
- Next, have one partner sort the buttons and the other partner guess the sorting rule.
- Pairs may share discoveries with the whole class.

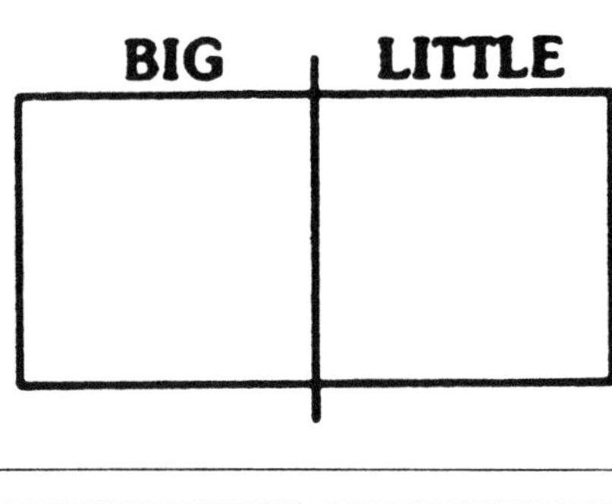

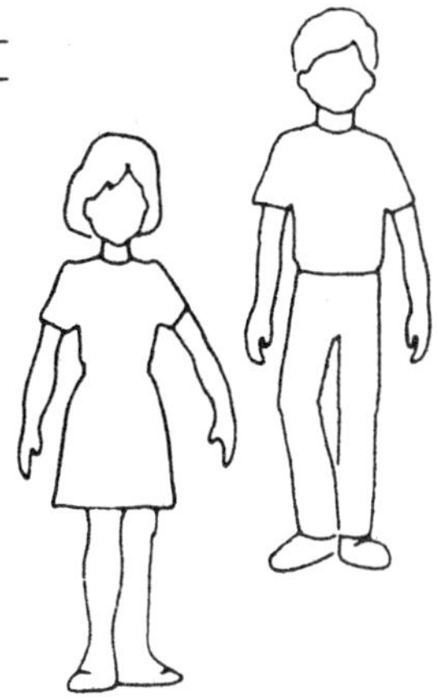

Family Logic

December

Grade Level: Primary

TASK ANALYSIS: 1 – Sort and classify objects by color
7 – Sort and classify objects by 2 or more attributes
9 – Put items into logical sequence
10 – Demonstrate the concept of all/some/none

MATERIALS: Data sheets, objects to symbolize family members; cubes, blocks, rods, whatever "junk" your class has available to manipulate, butcher paper, markers

ORGANIZATION: K – 3: Whole class activity, 1–2 days
Intergrate with Social Studies/Science
Cooperative groups: (later in the year, this may be done independently)

PROCEDURE:

- Xerox 1 data sheet per student.
- First day take home and gather data on family members eye color, hair color).
- Second day give each student a tub of colored manipulatives.
- Graph eye color.
- Graph hair color.
- Have them sort and classify by placing manipulatives in appropriate columns.
- Ask for sentences using all/some/none.

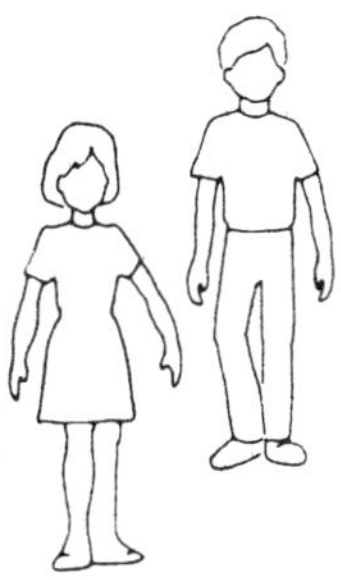

- On butcher paper graph, have students transfer data to class chart by placing cubes in appropriate column.
- Again ask students to give "math" sentences using all/some/none.
- Record responses on large chart paper.
- Examples of responses: "All the moms had brown hair." "Some of the moms had blue eyes." "None of the dads had red hair and blue eyes."

EXTENSION: Encourage children to use this activity at home with their families. Sort and classify by age, size, likes, favorite songs, etc.

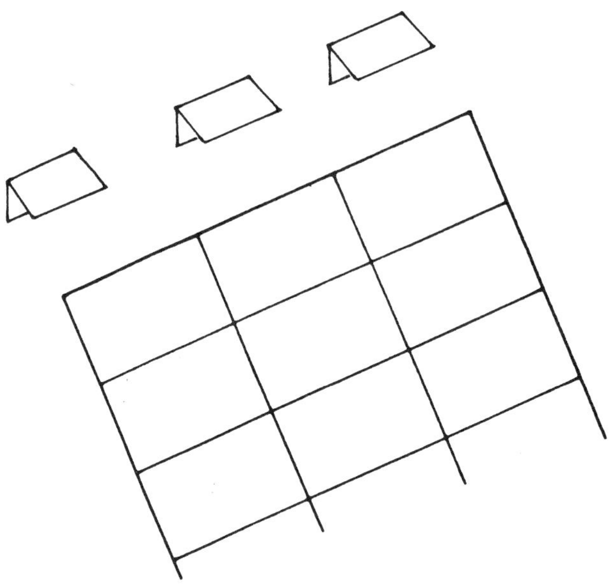

Ornaments

December

Grade Level: Primary

TASK ANALYSIS:
1 – Sort and classify objects by color
2 – Sort and classify objects by size
3 – Sort and classify objects by shape
7 – Sort and classify objects by two or more attributes

MATERIALS: A variety of ornaments brought by students and teacher large hoops or yarn to form circles

ORGANIZATION: Whole class activity
K–3: 20 to 30 minutes

PROCEDURE:
- Tell the history of how ornaments came into existence (see next page).
- Discuss the variety of ornaments that have been brought to school.
- Ask the students if someone has a way to sort the ornaments into the hoops.
- Have one of the students begin sorting according to his rule.
- Have the other students, one by one, place ornament in the hoops following the student's rule.
- The student says yes or no.
- Discuss the rule after all ornaments have been placed.
- Repeat the procedure (encourage the use of two or more attributes).

Ornaments From Many Lands

Over the centuries, people have used nuts, fruits, paper flowers, and the Christ child with angel wings to decorate Christmas trees. Every country has its special ornaments.

German ornament makers shipped their glass ornaments to the United States. In the United States, most ornaments were home-made. Strings of popcorn and cranberries along with paper chains and paper stars were hung on the trees. Some families hung popcorn balls.

In Lithuania people make delicate shapes and figures from straw. It signifies hope for a good harvest.

In Czechoslovakia, carefully painted eggs hang from the Christmas tree.

Polish people often hang angels and birds, especially peacocks, on their trees.

Japanese people cut and assemble pretty paper fans and lanterns.

Swedish people make brightly painted wooden ornaments and straw figures.

Today in America, many people use bright twinkling lights, bells, elves, reindeer, angels, stars, shiny tinsel, and toy horns. These seem to include symbols from many other lands.

Trim the Tree

December

Grade Level: Primary

TASK ANALYSIS: 8 – Identifies next item/step in a series
9 – Put items into logical sequence
12 – Demonstrate the concept of if/then

MATERIALS: Christmas tree, tree lights, children-made garlands (strung cranberries, popcorn, construction paper chains, etc.), children-made ornaments, tinsel

ORGANIZATION: Whole group instruction: K-3
You will need several days in order to make garlands and ornaments.
Actual lesson: 20-30 minutes

PROCEDURE:
- Several days in advance, brainstorm for items needed to decorate the class tree.
- Make some or all of the garlands and ornaments.
- The day of the tree-trimming, have the brainstormed items and children's garlands and ornaments together.
- Ask the children what should go on the tree first and why.
- Continue asking, "What comes next and why?"
- After the tree is trimmed, ask "what if..." questions: "What if we had put the tinsel on first?" and so on.

Snowman Logic

January

Grade Level: Primary

TASK ANALYSIS:
1 – Sort and classify objects by color
2 – Sort and classify objects by size
3 – Sort and classify objects by shape
4 – Sort and classify objects by thickness
7 – Sort and classify objects by two or more attributes
8 – Identifies next item/step in a series

MATERIALS: Marshmallows, toothpicks, markers, pictures of different snowmen, hoops or circles of yarn and colored miniature marshmallows

ORGANIZATION:
K – 1: Whole class activity, 2 days of 20 – 30 minutes
2 – 3: Cooperative groups

PROCEDURE:
- Brainstorm components of a snowman.
- Discuss how a snowman could be made using the above mentioned materials.
- Have students make snowmen.
- Ask students to explain the steps used in making their snowmen.
- Have students sort all the snowmen by attributes in hoops on the floor.
- Have students support their sorting by sharing their rule.

Attribute Sort

January

Grade Level: Primary

TASK ANALYSIS:
1 – Sort and classify objects by color
2 – Sort and classify objects by size
3 – Sort and classify objects by shape
4 – Sort and classify by thickness
7 – Sort and classify by 2 or more attributes

MATERIALS: Attribute blocks, sorting mats

ORGANIZATION: Cooperative groups, do activity over 2 –3 days
K – 1: 20 to 30 minutes; 2 – 3: 40 to 60 minutes

PROCEDURE:
- Introduce students to the attribute blocks: size, color, shape and thickness.
- Have students do several sorts beginning with color.
- Keep the color groupings.
- Next have students sort by thickness using the blocks in the color groups.
- Discuss classifications students found in their sorting.
- Next sort by shapes and keep the shape groupings.
- Have students sort by size using the blocks in the shape groups.
- Discuss classifications students discovered in their sorting.

Popcorn Logic

January

Grade Level: Primary

TASK ANALYSIS: 8 – Identifies next item/step in a series
12 – Demonstrate the concept of if/then

MATERIALS: Popcorn, popcorn popper, salt, butter, graph

ORGANIZATION: Whole group activity
K-3: 20-30 minutes

PROCEDURE:
- Brainstorm for the items needed to make popcorn.
- "What do we need to do first? Why?"
- "What comes next? Why?"
- Continue asking such questions until corn is popped.
- Graph the kinds of popcorn children like best. (See Hands On Probability, Statistics and Graphing).
- After the popcorn is made, discuss what was needed to make the corn pop.
- "What would have happened if we had not had heat?"
- Eat and enjoy.

Junk and Other Sorts

January

Grade Level: Primary

TASK ANALYSIS: 1 – Sort and classify objects by color
2 – Sort and classify objects by size
3 – Sort and classify objects by shape
7 – Sort and classify objects by two or more attributes

MATERIALS: Tubs of manipulatives (store bought or collected 'junk'), sorting mats, and recording sheets

ORGANIZATION: K – 1 cooperative groups of 2 to 4: 20 to 30 minutes
2 – 3 cooperative groups of 2 to 6: 20 to 30 minutes

PROCEDURE:
- Form cooperative groups.
- Get sorting mat and choose a tub of manipulatives.
- Have students sort using color attribute; record data.
- Sort using size attribute; record data.
- Sort using shape attribute; record data.
- Sort using two or more attributes; record data.
- Have each group explain one of their findings.

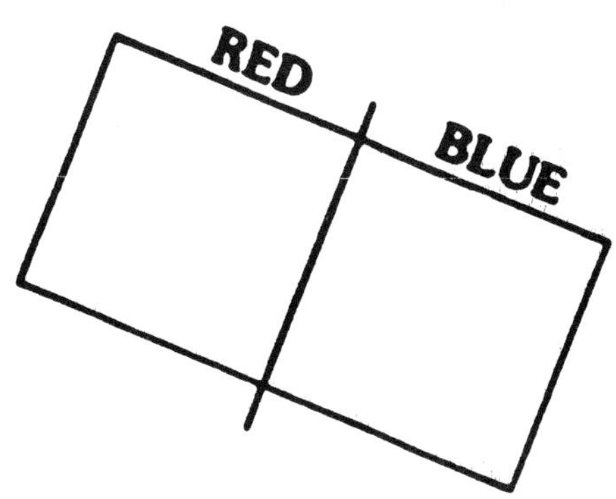

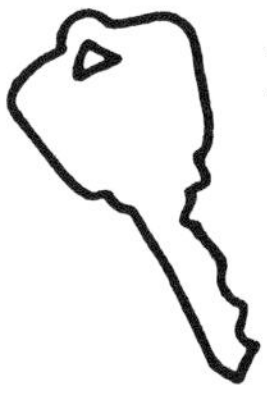

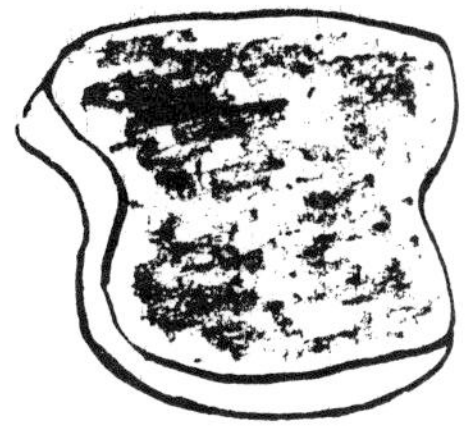

Cinnamon Bread Logic

February

Grade Level: Primary

TASK ANALYSIS: 8 – Identifies next item/step in a series
9 – Put items in to logical sequence
10 – Demonstrate the concept of all/some/none

MATERIALS: Several kinds of bread (wheat/rye/white), cinnamon, butter, toaster, spreading knives, butcher paper and markers

ORGANIZATION: K – 1: whole class activity, 20 –30 minutes
2 – 3: cooperative groups of 4–6 students, 30 – 50 minutes

PROCEDURE:
- Brainstorm to create the cinnamon toast recipe. Record.
- Students gather supplies and place in logical sequence on work area.
- Discuss students rationale for the given sequence.
- Have students prepare cinnamon toast according to their "recipe."
- Discuss results of the project.

EXTENSION: We find this exercise is a tremendous way to encourage the awareness of logic skills in your students throughout the year. Have students follow this procedure for any class function where refreshments may be involved (birthdays, holidays, social studies, open house, etc.).

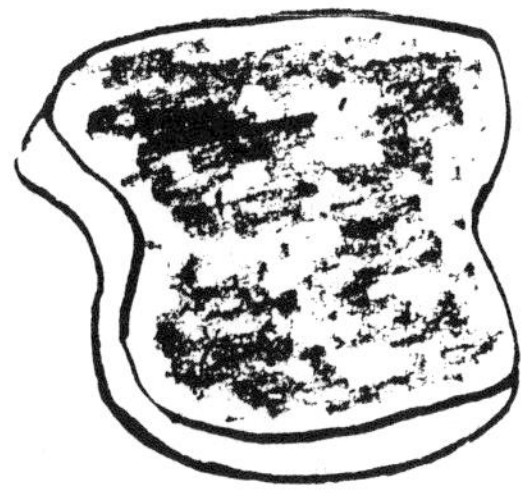

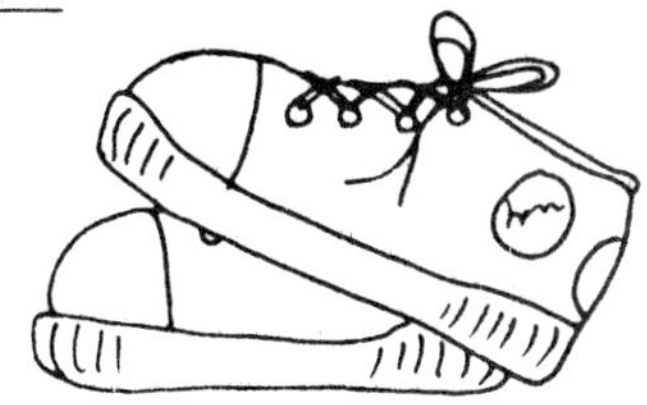

What Goes on What?

February

Grade Level: Primary

TASK ANALYSIS: 6 – Sort and classify objects by components

MATERIALS: Items that go on other items: pens with caps, pots with lids, hats, tables, tablecloths, lamps, shades, candles, matches, doll bed, blankets, shoes, socks; be creative! Hoops or yarn to make circles on the floor. 'Rule slips' and envelopes.

ORGANIZATION: Whole class activity K – 3: 20 –30 minutes
Suggestion: 2 –3 cooperative groups following whole class introduction

PROCEDURE:
- Place objects in center of circle.
- Have students brainstorm which items belong together.
- Teacher models placement of items in the hoops until a student sees the sorting rule.
- Have student place next item into hoop and teacher either says yes that fits my rule or no that does not fit my rule.
- Continue this process until several students understand the sorting rule.
- Ask students to explain conclusion. "How do the different parts of items relate to one another?" "May they be sorted differently?"
- Second and third graders may work in cooperative groups of 4–5 students to do the following lesson using similar items.

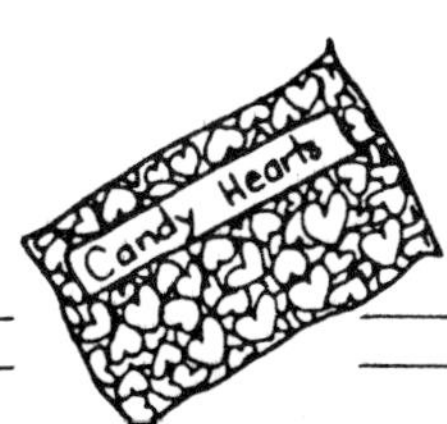

- Quietly (so other groups may not hear them) each group needs to decide on the "sorting rule" their group will use.
- Record on rule slip and place in "rule" envelope.
- Each group then sorts according to its "rule."
- When each group has written its rule and sorted its materials the coop groups begin to move through all the other groups' solutions.
- Each group determines the rule and records it on a rule slip then places it in the "rule envelope."
- Each group moves on to the next solution.
- Continue process until all solutions have been observed.
- When students return to their original place, they may open the envelope and see how many guessed the sorting rule.
- As whole class discuss the solutions.

Mixed-up Hearts

February

Grade Level: Primary

TASK ANALYSIS:
1 – Sort and classify objects by color
2 – Sort and classify objects by size
4 – Sort and classify objects by thickness
7 – Sort and classify objects by 2 or more attributes
10 – Demonstrate the concept of all/some/none

MATERIALS: Small candy hearts, large candy hearts, red-hot hearts, small cups (1 for each child or 1 for every 2 children), sorting sheets

ORGANIZATION: Individual or partner activity
K – 3: 20 minutes

PROCEDURE:
- Give each child (or pair) a cupful of candy hearts and a sorting sheet.
- Suggest the ways to sort and classify (T.A.'s 1,2,4,7).
- Have children explain their "sorts" and why.
- "Is there another way you can sort and classify the hearts?"
- Explain.

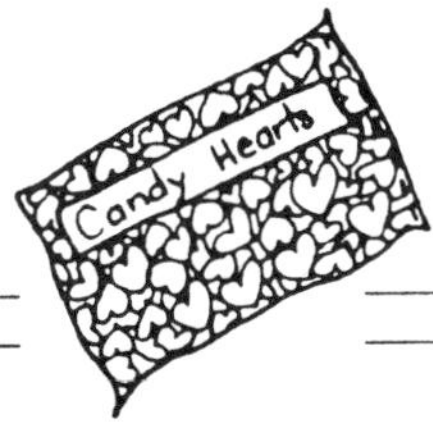

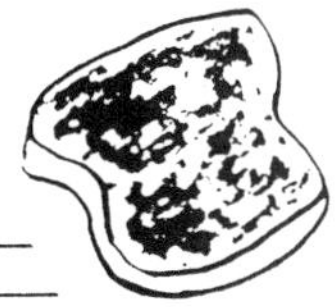

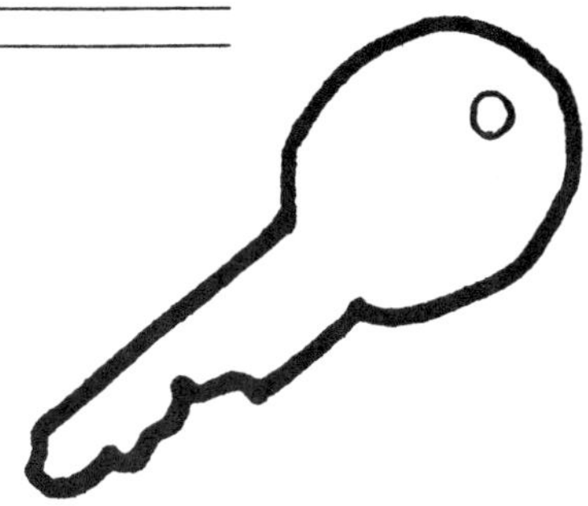

Keys to Logic

February

Grade Level: Primary

TASK ANALYSIS: 1 – Sort and classify objects by color
2 – Sort and classify objects by size
3 – Sort and classify objects by shape
4 – Sort and classify objects by thickness
7 – Sort and classify objects by 2 or more attributes
10 – Demonstrates the concept of all/some/none

MATERIALS: Variety of keys, sorting mats, hoops or yarn to make circles, floor graph

ORGANIZATION: Whole group activity; K – 3: 20 to 30 minutes

PROCEDURE:
- Give students a minimum of three keys and have them become familiar with the keys.
- Have students describe the attributes of each key.
- Have students place keys on the floor graph or hoops based on their chosen "rule".
- Discuss the different "rules".
- Ask students to make all/some/none/ statements based on the key graph.

EXTENSION: Students may choose to graph their data.

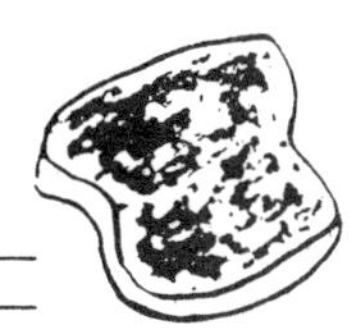

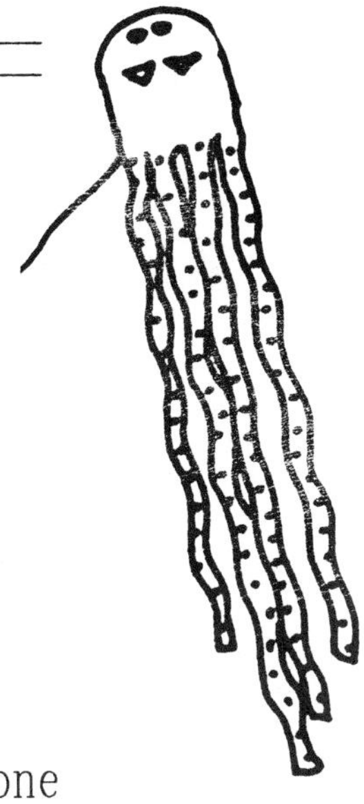

Make a Kite

March

Grade Level: Primary

TASK ANALYSIS: 6 – Sort and classify objects by components
8 – Identifies next item/step in a series
9 – Put items into logical sequence
10 – Demonstrate the concept of all/some/none

MATERIALS: Different materials needed to make a kite (paper, string, fabric, sticks, plastic); paper for recording sequence, pencils/crayons, butcher paper, headings.

ORGANIZATION: Whole class activity for K – 1
Small cooperative groups for 2 –3
If kites are to be made by each student, this may require several days.

PROCEDURE:
- Brainstorm student ideas of how kites stay aloft.
- Discuss the variety of materials that may be used in kite making.
- Prepare, as a group, steps to follow in making kites.
- Students make kites.
- Share results of kite making efforts.
- Graph the kites under the headings: Did Fly and Did Not Fly.
- Discuss the component parts of the kites under each heading.
- Possible student observations: "All kites that flew had tails. All kites that did not fly, had no tails. Some kites with tails did not fly."
- Students develop the ability to see relationships which give them the steps to reconstruct their kites for successful flight.

- This lesson may be integrated with a "Wind" unit and continue for several days.

Money Logic

March

Grade Level: Primary

TASK ANALYSIS:
1 – Sort and classify objects by color
2 – Sort and classify objects by size
3 – Sort and classify objects by shapes
4 – Sort and classify objects by thickness
5 – Sort and classify objects by use
8 – Identifies next item/step in a series
9 – Put items into logical sequence
10 – Demonstrate the concept of all/some/none
12 – Demonstrate the concept of if/then

MATERIALS: Several coins of all denominations and dollar bills: pennies, nickles, dimes, quarters, half dollars, dollars, butcher paper, markers

ORGANIZATION: Partners
K – 3: 20–30 minutes

PROCEDURE:
- Have the children work in partners.
- Give each pair coins and dollar bills.
- Have one child in each pair sort to one of the above tasks.
- Have the other child guess (in a sentence) how the money was sorted.
- Have partners trade "jobs" for the next sorting task.
- Have partners work together to put money into sequence (largest coin to smallest coin; smallest value to largest value; etc.) and explain their sequence.
- As a whole group, have the children classify by attributes as to all/some/none. Record on butcher paper.
- As a whole group, have the children put coins together to equal other coins (2 dimes and 1 nickle equals one quarter).
- Explain.

Dyed Easter Eggs

March

Grade Level: Primary

TASK ANALYSIS: 8 – Identifies next item/step in a series
9 – Put items into logical sequence
10 – Demonstrate the concept of all/some/none
12 – Demonstrate the concept of if/then

MATERIALS: Eggs; egg dye; bowls; spoons; markers; towels; pots/pans; water; butcher paper

ORGANIZATION: Whole class activity
K – 3: We suggest one day to dye the eggs and another day in which to discuss and draw conclusions.

PROCEDURE:
- Follow the directions on the package for dying the eggs.
- Have students gather the needed utensils and place in logical order of use.
- Have students prepare and dye the eggs step by step.

What's Related?

April

Grade Level: Primary

TASK ANALYSIS: 5 – Sort and classify objects by use
13 – Identify relationship among objects

MATERIALS: Several items from categories of clothing, school supplies, snack foods; furniture items; toys; be creative

ORGANIZATION: Whole class activity
K – 1: 20 to 30 minutes
2 – 3: 40 to 50 minutes

PROCEDURE:
- Place all the items on the floor area.
- Gather students around the items to be discussed.
- Identify each item by its use.
- Have children explain how the items relate to one another.

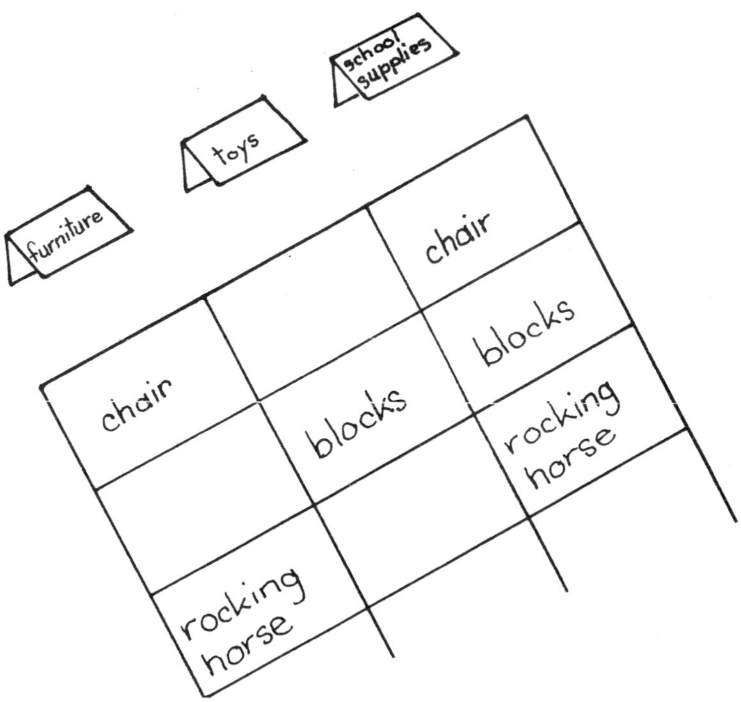

Peanut Butter and Jelly Sandwiches

April

Grade Level: Primary

TASK ANALYSIS: 8 – Identifies next item/step in a series
9 – Put items in logical sequence
12 – Demonstrate the concept of if/then

MATERIALS: Peanuts, blender, bread, knife, jelly, butter, jar of peanut butter (all in a large bag)

ORGANIZATION: Whole class activity
K – 2: 20 to 30 minutes

PROCEDURE:

- Brainstorm for things needed to make peanut butter and jelly sandwiches.
- Have items ready when children name the item.
- Have the children put the items into sequence: "What comes first, second, etc.; why?"
- Before doing the next part of this lesson, you may want to do a probability lesson. See Hands On Statistics, Probability and Graphing.
- "If we didn't have jars of ready-made peanut butter, what would we need to make it?" Brainstorm for items. (Have these ready.)
- Put the above items into sequence.
- Have children explain the sequence.
- "What do we have to do with the peanuts?" "Why?"
- "What do we do next?" "Why?"
- Make peanut butter and jelly sandwiches (Use the first sequence generated for making the sandwiches).

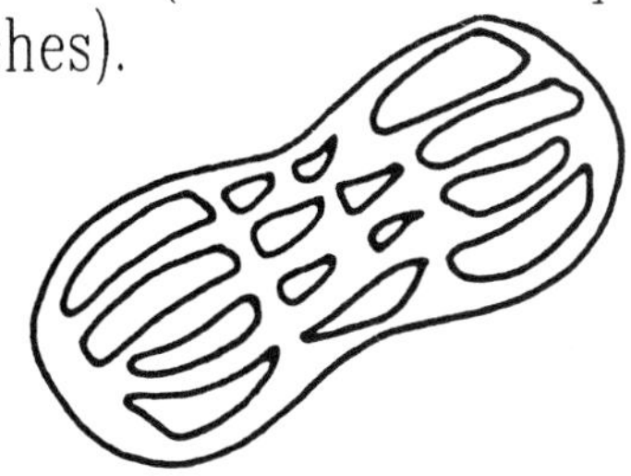

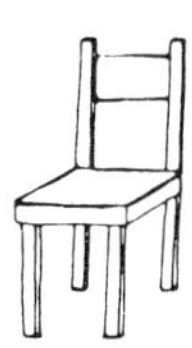

Clancy Clock Parts

April

Grade Level: Primary

TASK ANALYSIS:
6 – Sort and classify objects by components
8 – Identifies next item/step in a series
11 – Demonstrates the concept of or/and/not
12 – Demonstrates the concept of if/then
13 – Identify relationship among objects

MATERIALS: Items that show their component parts or that can be taken apart by components.: clock, train, car, record player, tape player, graphs, markers

ORGANIZATION: Cooperative groups
K – 1: 20 – 30 minutes
2 – 3: 30 – 40 minutes
Whole class activity (another session)

PROCEDURE:
- Have cooperative groups choose item that they will "figuratively" take apart.
- Sort and classify component parts.
- As a whole class have all cooperative groups sort and classify component parts.
- Discuss relationships of component parts.
- Have cooperative groups "figuratively" reconstruct the item explaining the sequence used.

May Baskets

May

Grade Level: Primary

TASK ANALYSIS: 1 – Sort and classify by color
2 – Sort and classify by size
3 – Sort and classify by shape

MATERIALS: Different shaped baskets and containers and/or your favorite paper pattern basket; flowers (real or artificial), ribbons, construction paper, and any other "junk" to decorate May baskets, floor graph

ORGANIZATION: K – 1: Whole group activity
2 – 3: Cooperative groups
K – 3: Will all need one work period to create the May baskets and one work period to discuss data and draw conclusions.

PROCEDURE:
- Have children choose materials they will use to make May baskets.
- Have children construct baskets.
- Sort/classify on floor graph by size, color, and shape.
- Discuss and interpret data.

Plants Logic

May

Grade Level: Primary

TASK ANALYSIS: 5 – Sort and classify by use
10 – Demonstrate the concept of all/some/none

MATERIALS: A collection of plants including vegetables, fruits, annuals, perennials, house plants, trees

ORGANIZATION: Whole group activity
K – 3: 20 – 30 minutes

PROCEDURE:
- As a whole group have the children sort the plants.
- Explain their classifications.
- Have the children classify the plants by attributes using all/some/none.
- Explain.

Making Ice Cream

May

Grade Level: Primary

TASK ANALYSIS:
2 – Sort and classify objects by size
3 – Sort and classify objects by shope
7 – Sort and classify objects by 2 or more attributes
11 – Demonstrate the concept of or/and/not
12 – Demonstrate the concept of if/then

MATERIALS: Ingredients for making ice cream (see Hands on Probability, Graphing and Statistics, pg. 77); different shaped cones; several ice cream scoops; butcher paper; marking pens; headings

ORGANIZATION: Whole class activity
K – 3: 20 – 30 minutes

PROCEDURE:
- Make ice cream.
- Each child prepares an ice cream cone (have all shapes of cones available and several ice cream scoopers).
- One cooperative group will sort and classify by scoops, another group by shape of cones; another group by 2 or more attributes.
- As a whole class activity, have cooperative groups share data.
- Children create graphs from the data.
- Ask questions or/and/not; if/then.

Pack Your Bags

June

Grade Level: Primary

TASK ANALYSIS: 2 – Sort and classify objects by size
5 – Sort and classify objects by use
10 – Demonstrate the concept of all/some/none
13 – Demonstrate the concept of if/then

MATERIALS: Suitcases of various sizes, a backpack, various items that could be packed determined by what kinds of "trips" you are going to take (big and small)

ORGANIZATION: Whole class activity
K – 3: Several days

PROCEDURE:
- "We're going on a trip!" Set some guidelines: "It will be hot," "We'll be gone three nights," etc.
- Brainstorm for things that will be needed on the trip.
- Have several items gathered for the children to sort and classify.
- "Which items will we select to use on our trip?"
- Have the children select a suitcase. "Will these items fit in the suitcase?" "Why?" "Why not?"
- The next day, go on another "trip" and repeat procedure.
- Another day, put all the items in their appropriate suitcases, and ask the children if some/all/none of the items could be used on both trips. Why? Why not?

EXTENSION: Children may graph their data.

Let's Relax

June

Grade Level: Primary

TASK ANALYSIS: 13 – Identify relationship among objects

MATERIALS: Pictures or actual objects the children have brainstormed

ORGANIZATION: Whole class activity
K – 3: 20–30 minutes, 2-day activity

PROCEDURE:
- First day: brainstorm for items needed for camping, swimming, and reading (or any other subjects).
- Have children bring pictures from home depicting these ideas or have them bring the real objects.
- Second day: have the children classify the brainstormed objects and tell how objects relate to one another.
- "Why does a book relate to camping?"
- "Can you swim while you camp?" "Why?" "Why not?"

14 Identifies relationships between two objects according to relative attributes

As Easy as ABC
Grade Level: Middle

MATERIALS: Newspaper or magazines cut outs of large alphabet letters, glue, art paper

ORGANIZATION: Groups of two

PROCEDURE: This is an early lesson in identifying relationships among objects. Begin by giving students a variety of magazines or newpapers from which to cut out large alphabet letters. They need only know that eventually they will be gluing these letters on to construction paper.

Once each student has collected fifteen to twenty-five letters, hand out two sheets of construction paper to each child. Tell students that their task is to divide their letters into two different attribute groups. Examples of these groups may include: letters that use only straight lines (A, M, N ,H,), letters that are curved shaped (C, G. O) , letters that use curved and straight lines (B, D, R), or upper, versus lower case letters. Many students may wish to arrive upon their own methods of grouping.

Have students glue all of one "attribute" letter type on one page, and the second type on the second page. Rather than having students explain how they divided the letters, you may wish to post the "collages" and have classmates form conclusions as to each student's method.

14	Identifies relationships between two objects according to relative attributes

A Missing Link

Grade Level: Middle

MATERIALS: Marking pens, word cards, for listing attributes, various items for display (candle, drinking glass, etc.)

ORGANIZATION: Groups of two, three, or four students

PROCEDURE: In this activity, students generate two attributes of a given item thus creating an "attribute chain" (see below). By then changing one of the attributes, students try to identify a new item which matches the given attributes.

Begin by having the class identify two attributes of a selected item (e.g. a candle). Prompt this response by listing several possible attribute categories on the board (size, color, shape, uses, component parts, or texture -- to name a few).

Using the candle as an example, students might generate attributes such as "white" and "cylindrical." Ask students if they can think of other objects which match these attributes.

Given this chain (below), ask students to change one attribute and then to try to identify an item which matches this new "chain." At this point you might divide students into groups to work on generating a group chain. Groups will then make replacements. You might also ask student to do this individually or with the help of parents as a homework assignment.

ATTRIBUTES	ITEM DESCRIBED
WHITE/CYLINDRICAL	CANDLE
RED/CYLINDRICAL	FIRECRACKER
RED/ROUND	APPLE
YELLOW/ROUND	LEMON

ETC.

14	Identifies relationships between two objects according to relative attributes

Pirate Treasure

Grade Level: Middle

MATERIALS: Boxes, various "junk" items from the home, enough for one box per group

ORGANIZATION: Groups of two, three, or four

PROCEDURE: Tell the students that they are going to sort the items in their boxes by attributes. They must name the attribute by which they are sorting and be ready to justify their reasoning.

The teacher should begin by choosing a box and dumping its contents out on a table. Each item should be held up and displayed for the class to see. After all the items have been displayed the teacher should name an attribute and begin sorting by it -- making deliberate mistakes to test students' understanding. Once all items are sorted by this given attribute, call on the class to help name another attribute for sorting and re-sort by this new attribute.

Examples of attributes would include: size, color, shape, use, texture, or component parts, etc. Discuss these and other possible attributes.

After the sorting has been done by the teacher with the whole class participating, divide the students into groups and have them proceed to sort materials in their boxes by a group-determined attribute. When they have finished sorting, have them label each pile with a descriptive adjective and allow them to circulate to examine other groups' sorting and naming of their attributes.

When the class has had time to examine each other's work, hold a general discussion about the different types of attributes chosen and sorting process.

14	Identifies relationships between two objects according to relative attributes

To Share and Share Not

Grade Level: Middle

MATERIALS: Sets of items with similarities and differences (apples/grapes; leaves/flowers; scissors/pliers)

ORGANIZATION: Groups of two, three, or four

PROCEDURE: Begin this lesson with the teacher holding up one item (apple) and asking students to tell various attributes of the item -- do not write them down. Hold up a second item (grapes) and repeat the procedure of identifying attributes.

Ask students to then decide if there are any shared attributes of the two items and write these on the board. Ask students if there is any way that they could identify one item versus the other by identifying only the shared attributes (not possible). Then list one or two different (non-shared) attributes of the two items. Once again, ask if they can identify which item is which by these differences (they can).

Divide student in teams of two, three, or four and give each team two items. Ask them to write one or two shared attributes and one or two non-shared attributes of the two items. Upon completion, have each group share their statements with class members.

The closure of this lesson is to elicit the statement from the class that items can only be differentiated by their DIFFERENCES. Similarities allow things to be grouped.

As an extension, you might ask students to create Venn Diagrams which identify the unique and shared attributes of the two items they have compared.

14	Identifies relationships between two objects according to relative attributes

A Fool's Tools

Grade Level: Middle/Upper

MATERIALS: "Tools/machines" pictured on next page (it may also be possible to obtain examples of various tools from local museums)

ORGANIZATION: Groups of two, three, or four students

PROCEDURE: Explain that by analyzing the component parts of an object, a logical conclusion can be reached as to its practical uses. For example, show the students a screwdriver and ask if it looks as if it would be used for sawing wood.

What qualities and component parts does a saw have that a screwdriver doesn't, which enable it to do the task of slicing wood? Does a hammer look as if it would be useful for sewing? Why not? Could a shovel be successfully used as a wrench? Why not? The point here is that by close examination of an object's component parts and qualities, logical conclusions can be made as to its overall purpose and use.

Pass out the copies of the pictures of tools and explain that as civilization has progressed, so has its technology. This has caused a change in the types of tools people use to accomplish various jobs. These are pictures of tools which were in fairly common use in the 1800's. The task for the students is to conclude the original use of these tools. Have them list the facts about the tools they can observe. This should make it easier for them to connect it with a use. Do not have them share their results.

After each group has finished their speculation, post their responses so students can compare findings. Discuss which ones are the "most logical," which ones are the closest to correct, and which ones are least logical. Explain that even if a group did not accurately identify the tool's use, the answer should still be considered correct if their explanation was arrived upon by logical reasoning.

An interesting and fun extension (possibly for homework or free classtime) would be to have students draw pictures of tools of the future (one hundred years from today).

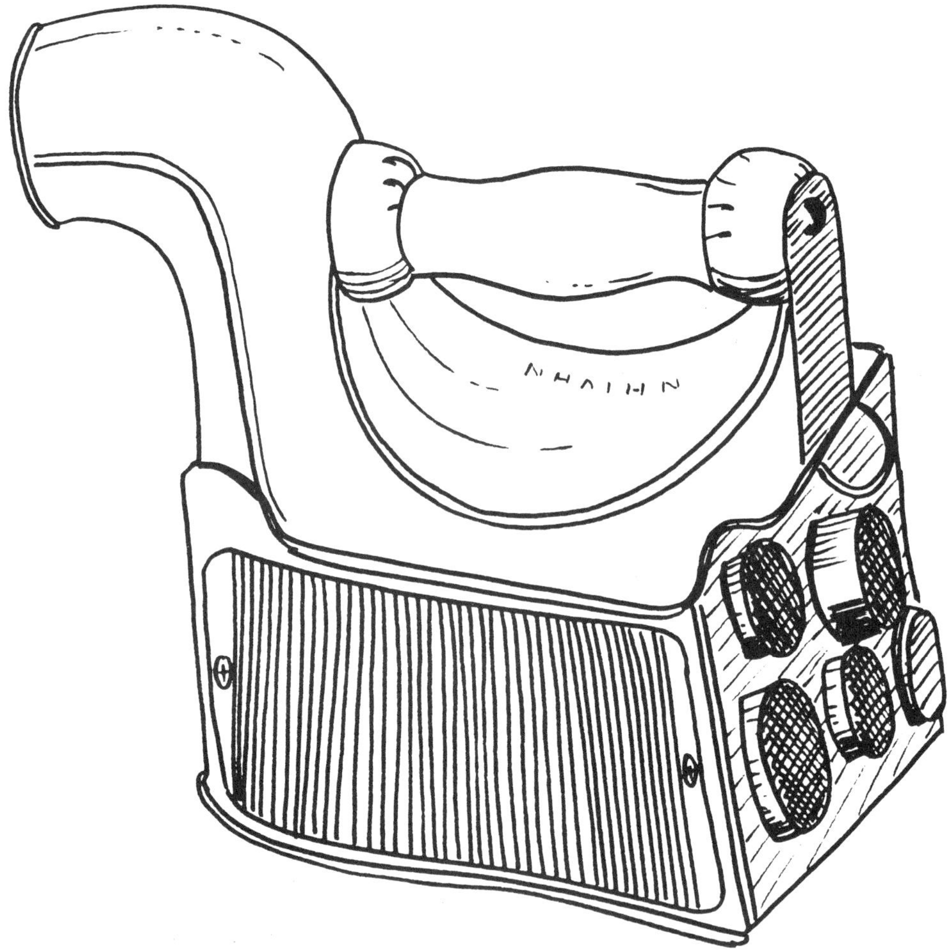

A).

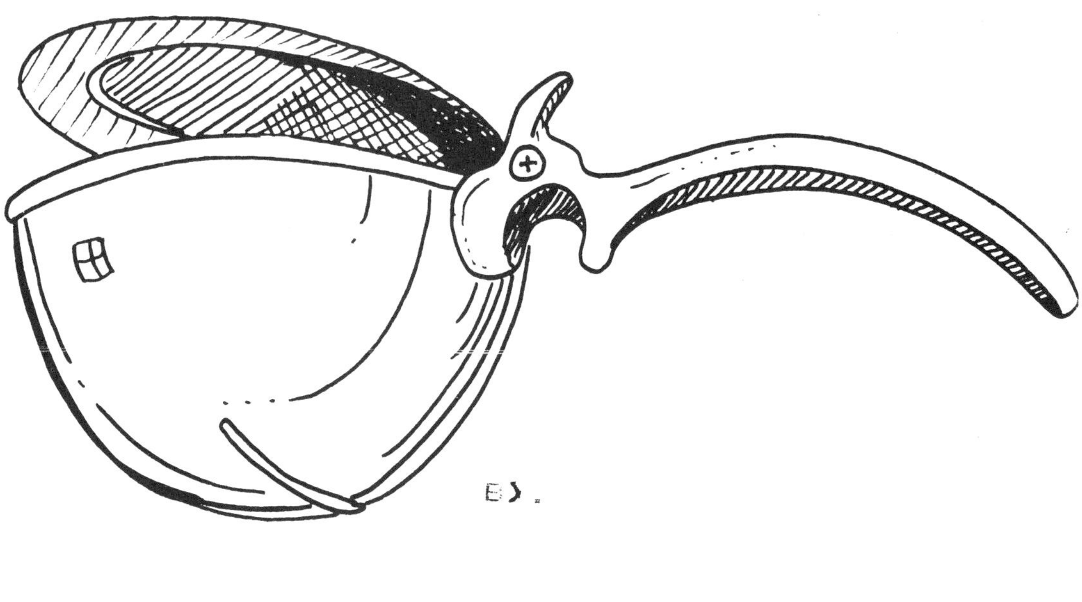

B).

A). Iron B). Silent Butler/Ash Caddy

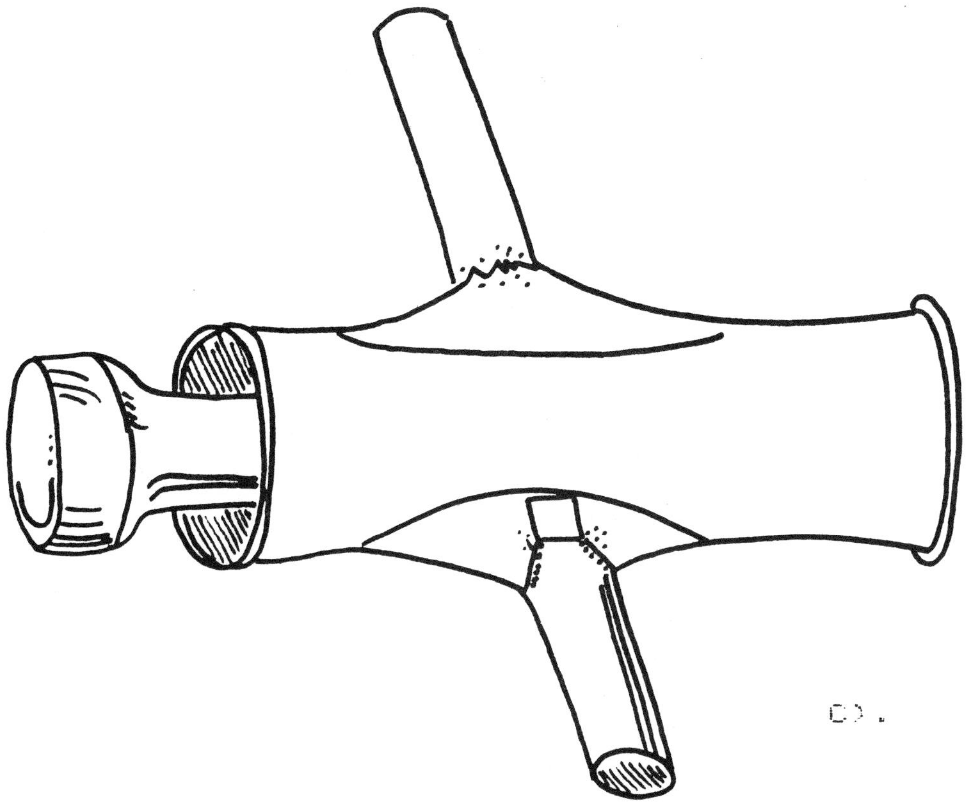

C).

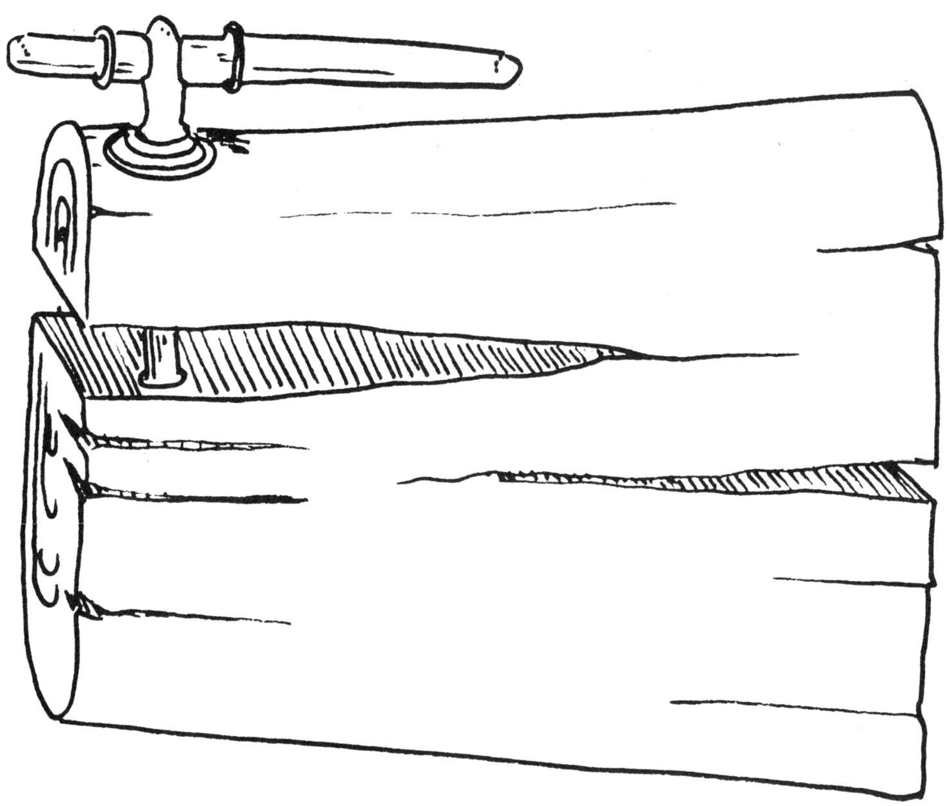

D).

C). Sausage Stuffer D). Horse-collar Stretcher

E).

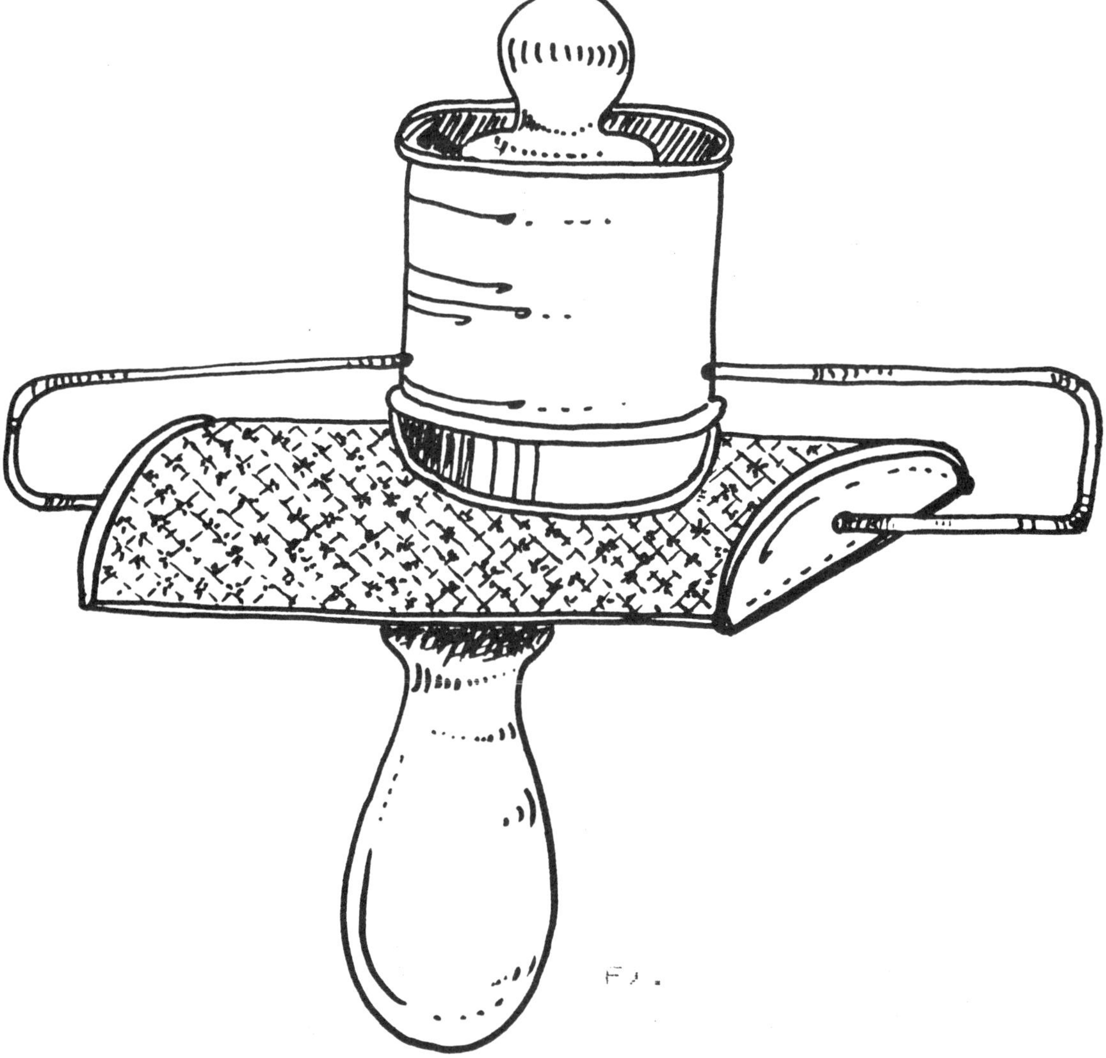

F).

E). Spittoon F). Nutmeg Grater

14	Identifies relationships between two objects according to relative attributes

Music, Music, Music

Grade Level: Middle

MATERIALS: Various songs (listed on next page), construction paper, glue

ORGANIZATION: Whole class activity

PROCEDURE: Tell the students that they are going to listen to the beat (rhythm) of two songs and attempt to imitate the beat by clapping or walking. They will then record the beat by organizing colored paper strips (see below).

Play a record or tape, or have the children sing two songs while the teacher models by clapping the beat. After each song have the children imitate the beat as a whole class (by clapping). You may also wish to select one or two students to see if they can "walk" around in a circle to this clapped rhythm (this will be difficult for the children).

Ask students if there is any way they can change the sound of claps or the movement of walking into something they can see. Lead them to a suggestion of using a colored strips of paper to represent different length notes, e.g. red for eighth notes, yellow for quarter notes, blue for half notes, and black for whole notes (the songs listed below avoid syncopated or dotted note patterns). If you are not familiar with music notation, use a music book as a guide for creating a color pattern for each song.

Another solution to "pictorializing" the music is to use different lengths of paper for each note duration.

Sample notation for Yankee Doodle using patterns and colors.

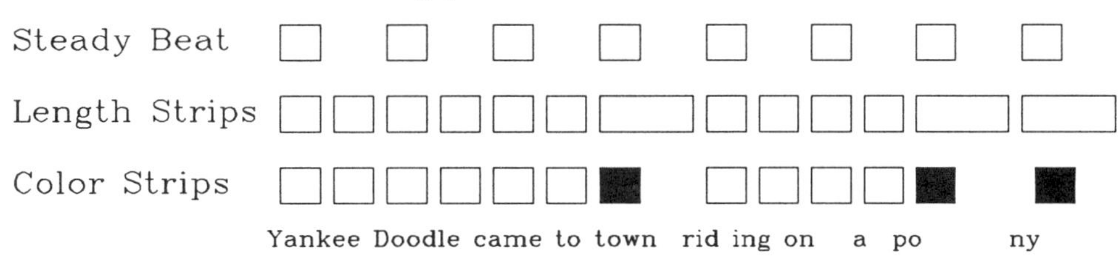

Getting to this point in the lesson has involved numerous logic skills along with a great deal of time. Continue with the lesson and have students generate a second picture of a song (as shown for Yankee Doodle). Once the second chart is complete, have the class discuss the similar attributes of the two songs.

Common attributes might include the same types of notes (colors or strip lengths), same groupings of notes, same organization of notes -- i.e. several "quick notes" followed by a long note.

An extension of this portion of the lesson is to have children analyze the up and down movement (pitch of the notes). You may wish to do a second chart showing this movement. On the charts below we demonstrate a variety of ways to "picture" songs.

While these activities are very sophisticated, your students will become very involved. Most children are inherently interested in music and when given the opportunity or explanation of how to describe and compare music, they will anxiously participate.

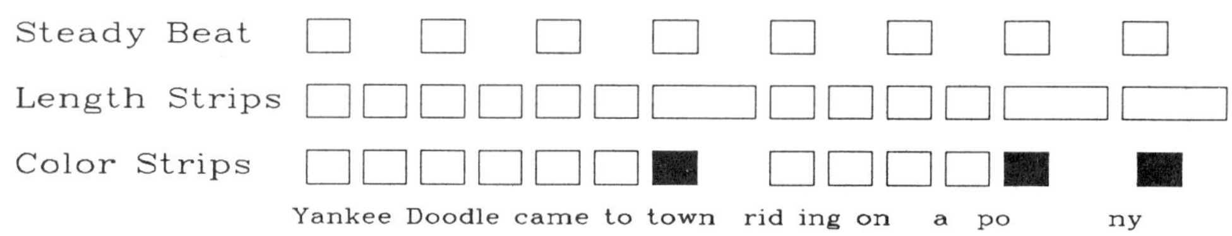

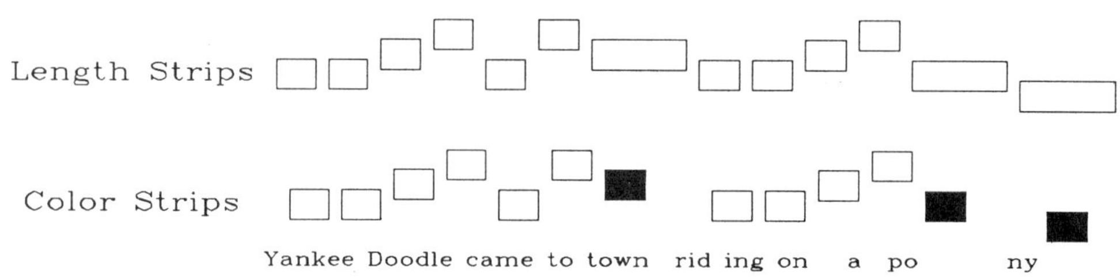

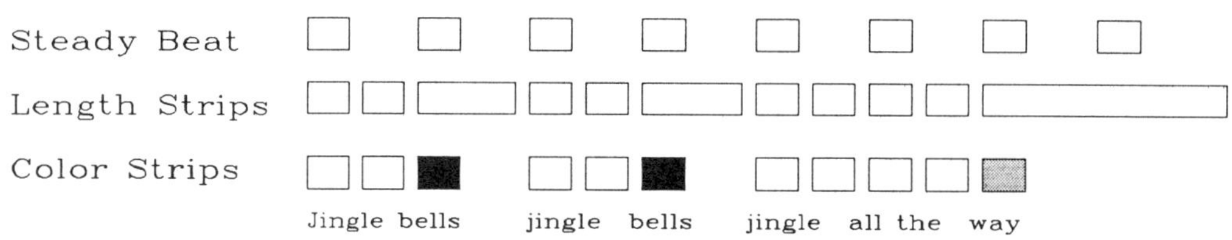

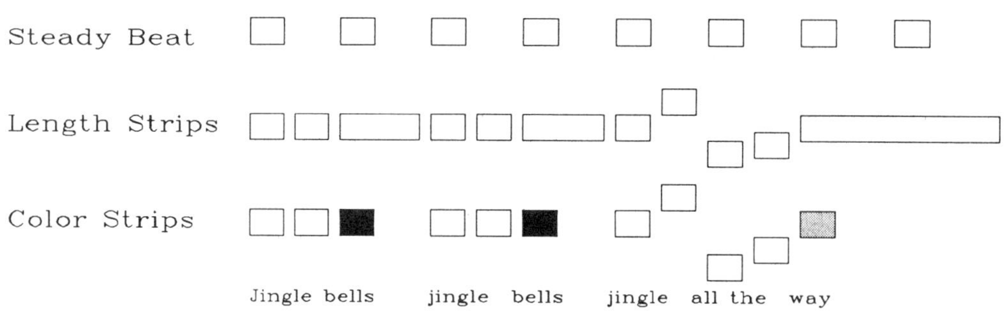

15	Identifies simple relationships among objects according to position in a pattern or sequence

Leaf Them in a Pattern

Grade Level: Middle

MATERIALS: Assortments of leaves collected by the students, a teacher prepared leaf pattern (see below)

ORGANIZATION: Groups of two

PROCEDURE: In this activity, the students will identify the relationships among a random group of leaves according to their position in a pattern. The teacher should glue several different types of leaves in a prearranged pattern on a sheet of paper.

Begin the lesson by holding up the teacher prepared sheet and asking students to identify various attributes of each leaf. Ask them to see if they can find any pattern. Ask them to explain the relationship among the leaves in the pattern (i.e. serrated, oval, slender, serrated, oval, slender).

Give each team of students a collection of ten to fifteen various leaves (or have them use their own collections). Have the students glue leaves on paper in their own patterns and exchange with other students, asking one another to explain the relationship among the leaves in these patterns.

As an extension, there is an obvious tie in to a science lesson on leaf identification or photosynthesis.

15	Identifies simple relationships among objects according to position in a pattern or sequence

What Goes Where?

Grade Level: Middle

MATERIALS: Sunday newspaper comic strips, glue, art paper, large envelopes

ORGANIZATION: Individual or groups of two

PROCEDURE: Begin by handing out a comic strip to each student or team. Have students cut each frame apart and place them in an envelope. Put the title of each comic strip on the outside of the envelope.

Then have students trade envelopes, arrange the comics in order by sequence, and glue them on paper. Next, have them trade the finished comics with classmates to check for "correct" sequence. Allow students an opportunity to challenge one another as to the order even if the layout appears to be correct. This justification of sequence involves logical reasoning for establishing pattern. Students need to learn to explain why things are as they are.

15	Identifies simple relationships among objects according to position in a pattern or sequence

Now That's an Order

Grade Level: Middle

MATERIALS: Story given below, glue, paper

ORGANIZATION: Individual or group of two

PROCEDURE: Begin by saying that you have a story for the students but that you dropped it on the way to class and they need to put it together before it can be read. Each sentence in the story should be cut apart and handed to the student without any pattern in the sentences. The students' task is to find the correct sequence of the story and paste it in the correct order. Have students read their stories aloud to the class to see if they are: 1) the same, 2) have a logical order, or 3) if they make sense.

THE BUNDLE OF STICKS

There once was a man who had four sons.
The father loved them very much, but they troubled him greatly.
For they were always fighting with one another.
Nothing the father said did any good.
"What can I do to show my boys how wrong it is to act this way?"
One day he called the boys to him and showed them a bundle of sticks.
"Which one of you, my sons, can break this bundle of sticks?" he asked.
All the boys tried in turn, but not one of them could do it.
Then the father untied the bundle and gave each boy a single stick.
"See if you can break that?" he said.
Of course, they could do it easily.
"My sons," the father said, "each of you alone is weak.
He is as easy to injure as one of the sticks.
But if you will be friends and stick together, you will be as strong as the bundle of sticks."

As an extension, you can use many fables or folk tales in this manner

15	Identifies simple relationships among objects according to position in a pattern or sequence

Four (Not) in a Row

Grade Level: Middle

MATERIALS: Four pieces each : split peas (green), dry beans (red), wheel macroni (tan), sugar cubes (white), or other similar items for each student, a grid of four squares

ORGANIZATION: Individually or groups of two

PROCEDURE: This problem requires the child to arrange the pieces in a four by four square so that there is one piece of each color and one piece of each shape in each row and in each column, and that no colors or shapes may be in the same row. While some students may find this very simple, many will have great difficulty in completing the task.

While it is not hard to establish an arrangement that meets either the color requirements or the shape requirements, it is hard to fulfill both requirements at the same time. Transposition of pieces will affect both the shape and color distribution, of course. It may be helpful to call attention to the number of colors in the diagonals or in each quadrant.

Ultimately, the goal is to have each child tell the strategy used in finding the solution. For students who solve the puzzle easily, have them experiment with a second or third solution or increase the difficulty by adding that the same color or the same shape cannot align diagonally (this in addition to not being in the same row).

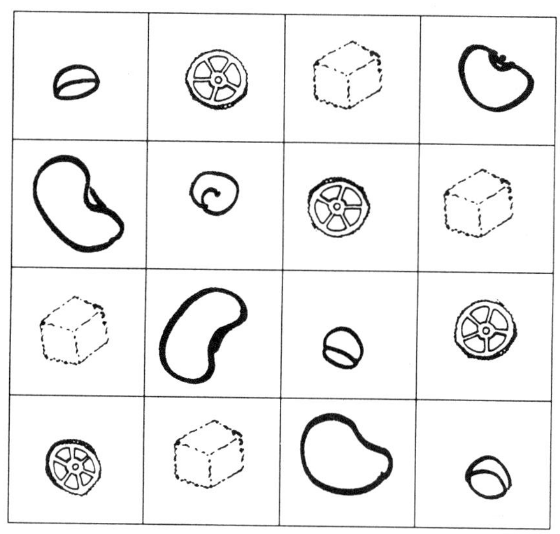

16	Uses the techniques of Venn Diagrams to represent relationships (sets, subsets, etc.)

Moving to the Bullseye

Grade Level: Middle

MATERIALS: Markers and tagboard name cards (tents), yarn or string for large classroom-sized circles

ORGANIZATION: Whole class

PROCEDURE: This is an early lesson in Venn Diagramming. You will need to make two concentric circles (use string) large enough to have standing room for most the students in the class. You may need to do this activity outside.

Direct students to stand around the large circle, not inside. Place a tagboard tent labeled UNIVERSAL SET on the floor outside the circles and explain to the students that everyone in the room is making up the universal set, simply because they are all part of the class.

Place the next tagboard sign labeled BIRTHDAYS in the outer circle and tell students who have a birthday to step into the first circle (this will obviously include everyone). Next, place the last tagboard signed labeled JUNE, JULY, AUGUST, AND SEPTEMBER, in the center circle and ask students who have birthdays in those months to step into the central circle.

Discuss this procedure with your students. Ask questions about why this sort of Venn Diagram representation might be useful. What generalizations can they make about a Venn Diagram with a circle (or circles) within a large circle? Have students propose other organizations which fit the patterns shown below.

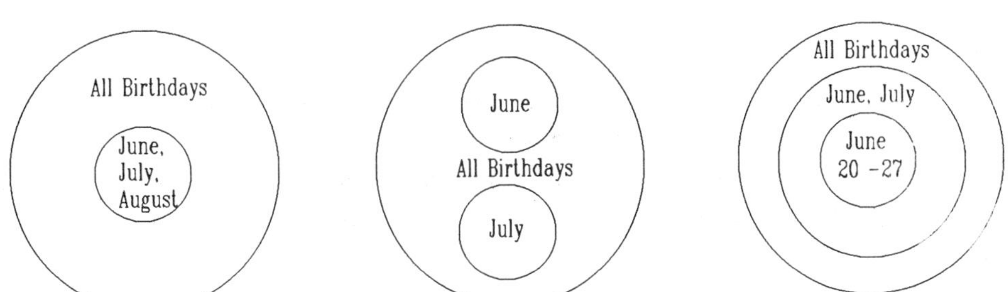

16	Uses the techniques of Venn Diagrams to represent relationships (sets, subsets, etc.)

Pass On the Pasta

Grade Level: Middle

MATERIALS: 6 to 8 different kinds of uncooked pasta (of as many colors and shapes as possible), string or charts for making Venn Diagrams

ORGANIZATION: Two or three large groups

PROCEDURE: This lesson is an early attempt at finding similar and dissimilar attributes between objects that share some common attributes. The goal is to encourage critical observations and to classify these observations.

Begin by dividing the class into two or three large groups, giving each group ten to twenty various pieces of uncooked pasta types. Ask the students to try to group the pasta types by some attribute, this might be by color, shape, thickness, etc. Given these instructions, some students will have two or three groupings, some may have seven or eight groupings.

Have students cut lengths of string and tie the ends together to circle each of their groups. The product will look like this:

Next, ask students to select two or three of their pasta groups and ask if there are any shared attributes between the them. Students should check for ways to overlap strings so that certain pasta types could be placed in two different circles (i.e. if students sorted by shape, they may find that colors may be shared by two groups). Have each student group experiment with overlapping the string circles and the pasta types. Students will discover a variety of configurations. Group members should explain their Venn Diagrams to other groups as they move around the room.

16	Uses the techniques of Venn Diagrams to represent relationships (sets, subsets, etc.)

A Feel for Venns

Grade Level: Middle

MATERIALS: Large collection of objects to be sorted by tactile attributes, string or charts for making Venn Diagrams, and bags for each group

ORGANIZATION: Groups of four to six

PROCEDURE: Brainstorm with students for words or ideas that express the way objects feel when we touch or feel them. Post these student wordbanks on the chalkboard or around the room for easy student reference.

Begin by explaining to students that they will be sorting objects according to the way they feel (no fair peeking). Remind students that they will be required to explain their classifications when this procedure ends.

Place eight to ten objects in a large shopping bag. On the chalkboard write the words, SMOOTH, FURRY, and ROUGH. Ask a student to come forward and feel in the bag to select one item. It should not be removed from the bag and classmates should not be told what the item is. Have the student go to the board and put a check next to the word SMOOTH, FURRY, or ROUGH. The student should not say what the item is but will need to remember what he/she selected. Continue this process, giving each child in class an opportunity to reach into the bag.

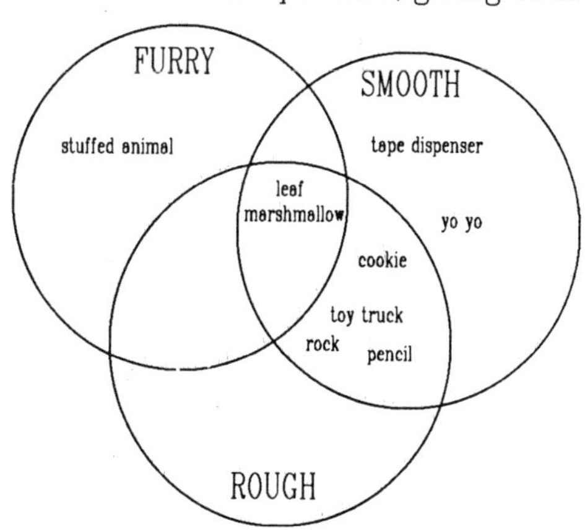

Next, draw large circles around each of the adjectives on the board and call on the students one by one to identify the item which they felt in the bag. Write the name of the item in the appropriate circle. You will find that several items will be classified in two of the circles. This is the basis of the discussion for shared attributes and Venn Diagrams. Hand out paper and markers, and have students use the information on the board to create Venn Diagrams which show the intersection of the SMOOTH, ROUGH, and FURRY attributes.

Once students have completed the task, make three large circles of string on the floor and pull the items one by one from the bag and have students place them in the correct location.

16	Uses the techniques of Venn Diagrams to represent relationships (sets, subsets, etc.)

Picture This Diagram

Grade Level: Middle/Upper

MATERIALS: Magazines, glue, scissors, and large paper (16" x 20") for making a collage

ORGANIZATION: This activity could be done as an individual or as group assignment

PROCEDURE: Each person or team is to pick a topic or theme for their collage. They are to choose and collect pictures and arrange them so that when someone else studies the collage they can find the evidence of a Venn Diagram even though there are no circles to delineate the boundaries. The placing of the pictures will be very important.

For explanation purposes, we will use an example of a collage of clothing. Begin by cutting out many pictures of items of clothing; pants, shirts, shoes, girls' vs. boys' clothing, hats, etc. At this point there is no need to worry about the concept of the collage. You might even have students perform this first step at home as a homework assignment.

When students have a collection, have them sort them by attribute. There are a variety of approaches they might take: color, purpose, male or female wearer, seasonal use, etc. Once students have completed a sort, have them lay the pictures out on a sheet of construction paper. The diagrams below show various possible constructions. Once students are satisfied with their layout, have them glue the pictures and display them for class discussion.

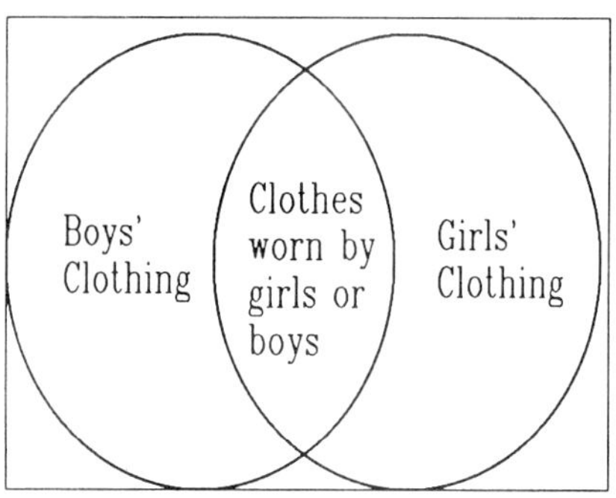

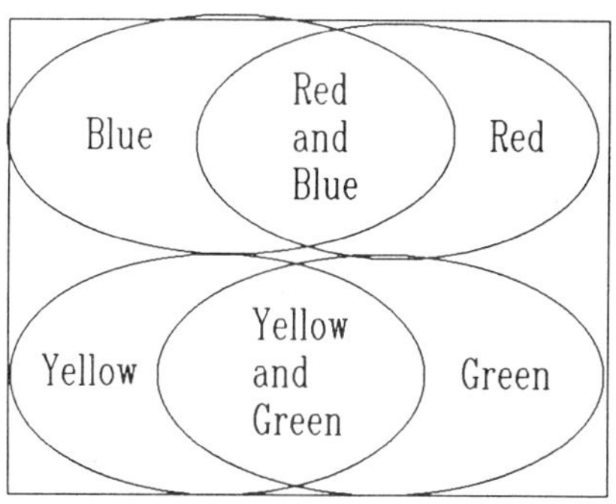

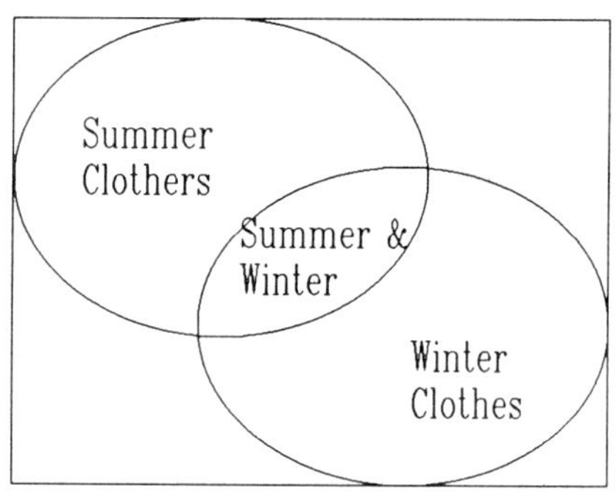

17	Demonstrates an understanding in using the words: ALL, SOME, and NONE

Two By Two

Grade Level: Middle

MATERIALS: Animal crackers (pictures of animals may also be used) for each cooperative group

ORGANIZATION: Cooperative groups of three, four, or five students

PROCEDURE: This lesson uses boxes of animal crackers (cookies) to provide students with materials for developing ALL, SOME, and NONE statements.

Divide the class into groups of three or four and give each group a box of animal crackers. Begin by having each student draw one animal from the box. Within the group, the students should create three statements, one for each of the key words...

ALL of our animals have four legs.
SOME of our animals live in the United States
NONE of our animals are birds

Continue this process, each time asking group members to create new ALL, SOME, and NONE statements.

Once all animal crackers have been selected, have groups make generalizations about the entire contents (i.e. – All animals in the box are mammals, some of the animals are carnivores, none of the animals can be kept as a pet). Have each group share their statements with the entire class.

As a fun extension, play an exception game in which students get to eat the selections which are exceptions to the rule. For example, player 1 might say, "All of the animals have hooves." If students have animals that do not have hooves, they get to eat those animals. Do this until each child can identify exceptions allowing him to eat his cookie selections.

17	Demonstrates an understanding in using the words: ALL, SOME, and NONE

Making it Stick

Grade Level: Middle

MATERIALS: Ten to fifteen household or classroom items in a bag and a magnet for each group

ORGANIZATION: Groups of three or four students

PROCEDURE: Begin the lesson by discussing what a magnet is and what it can do. You may wish to precede this lesson with a science lesson on magnetism.

Give each group a paper bag full of household or classroom items such as paper clips, chalk, clothespins, pencils, scissors, pennies, keys, paper, index cards, etc. Include 10 to 15 items in each bag.

As one group member holds the bag open, let another team member reach in the bag with the magnet without looking, and attempt to pick up an object. Remove any items that the magnet attracted from the bag. Let a second group member and then a third member repeat this procedure. At this point, have the group make statements about the bag contents. Appropriate conclusions might include:

ALL of the magnetic items have been removed.
SOME items in the bag are not metal.
NONE of the items left in the bag can be lifted by the magnet.

Have students generate numerous ALL, SOME, NONE statements and then check the bag to verify their deductions.

As a final step, have students sort the bag items by attribute, each time creating ALL, SOME, NONE statements (i.e., ALL items can be held in the hand, SOME of the items are made of wood, NONE of the items have a value of more than $.75).

17	Demonstrates an understanding in using the words: ALL, SOME, and NONE

All, Some, and None Numbers

Grade Level: Middle

MATERIALS: No special materials are necessary

ORGANIZATION: A whole class activity

PROCEDURE: In this lesson, students will create statements to answer questions about number sentences. There is some excellent review of the the greater than $>$, and less than $<$ symbols as well as some learning about algebraic expressions.

Begin the lesson by writing a simple sentence such as $3 + 5 = ?$ on the board. Ask students for the solution and extend this response to have students say that when an equal sign is used, there is generally only one solution. Now write $3 + ? = 8$ and ask the same questions and discuss the same conclusion.

Next, write $3 + 5 > ?$ and $3 + 5 < ?$ and ask students for the answer. You will find that student will have numerous responses. Let students know that with the use of ALL, SOME, and NONE, you can create statements that will include all possible answers. Ask one student to create a statement such as, "all numbers smaller than 10," and write this one the board. Using the number sentence, $3 + 5 > ?$, is it correct to say: ALL numbers smaller than 10 solve this equation?
None of the numbers smaller than 10 solve this equation?
Some numbers smaller than 10 solve the equation?

You will probably need to help students discover that the correct answer is "SOME." Ask students to think of a statement that would work with the words ALL (all numbers larger than 8) and NONE (none of the numbers smaller than 8). Continue this process with the $3 + 5 < ?$.

Once students have the idea, write three or four number equations on the board and let students generate a number sentence (e.g. numbers larger than 10), and try to place the correct AND, SOME, NONE word to make the sentence true.

As an extension, you might use statements such as, "all even (odd) numbers" and "all factors of..."

17	Demonstrates an understanding in using the words: ALL, SOME, and NONE

I Can Picture That!

Grade Level: Middle

MATERIALS: Magazine pictures of animals which fit into categories (e.g. zoo or farm animals)

ORGANIZATION: Groups of four or five

PROCEDURE: This is an early lesson in the concept of ALL, SOME or NONE.

Post on the chalkboard or overhead projector the phrases listed below and explain to students that by placing the words ALL, SOME, or NONE in front of each of these statements they will become true. Their goal will be to look at a set of animal pictures and decide which word to place in front of each statement.

Display four to six animal pictures for the students to view. After viewing pictures, have a student volunteer place an ALL, SOME, or NONE statement in front of the phrase "...of the animals are mammals." Discuss whether or not the student's response is logical. Continue this process with the other questions.

SAMPLE STATEMENTS

ALL ?

SOME ?

NONE ?

...of the animals are mammals
...of the animals eat meat
...have two eyes
...have two or three legs
...have brown fur
...of the animals are birds
...of the animals swim
...have wings

There are many possibilities for integration of this activity into a science unit on animals, plants, or earth science.

17	Demonstrates an understanding in using the words: ALL, SOME, and NONE

Sometimes It's All or Nothing

Grade Level: Middle/Upper

MATERIALS: Word cards, markers, word card rack for rearranging statements

ORGANIZATION: A whole class activity then dividing into small groups

PROCEDURE: This lesson involves the use of syllogisms in showing students the effects of the words ALL, SOME, NONE.

In preparation for the lesson, make a set of cards, which say ELM/TREE/LEAVES/GREEN, make another set of cards which say, ALL/SOME/NONE or NO (three of each card) and any other word cards as needed. Begin with the following syllogism in the word card rack.

IF Elms are trees
AND trees have leaves
THEN Elms have leaves

Call on a student to come forward and select one of the ALL, SOME, NONE cards and place it in front of one of the original cards (elm, trees). Ask students to note and discuss the new meaning of the syllogism.

Ask a second student to come forward and substitute another "new" card in the syllogism. Discuss the new meaning and continue with this process of adding (or removing) various new cards.

The value of this lesson is in the discussion and logical reasoning (arguments) which will ensue. Encourage small groups of students to write their own syllogisms. With practice, students will begin to see the effects that qualifiers such as: ALL, SOME, NONE, have on statements.

Other sets of words: Hershey, candy, chocolate, wrapper
zebra, black and white, stripes, animal
scissors, cutting, sharp, blades.

17	Demonstrates an understanding in using the words: ALL, SOME, and NONE

Verbigories

Grade Level: Middle/Upper

MATERIALS: Word cards, markers, chalkboard or overhead projector

ORGANIZATION: A whole class activity, later as cooperative groups

PROCEDURE: This lesson provides an excellent opportunity to integrate a language lesson into math. Students will need to have a working knowledge of a particular part of speech (we recommend using verbs or adjectives).

Begin by drawing a chart (as pictured) on the chalkboard.

ALL	SOME	NONE

Ask a student to randomly say a noun, let's use BIRD as an example. Students will now be generating verbs (or adjectives) which always, sometimes, or never describe birds. You may want to begin with providing some ideas, such as "waddle," "sing," "eat," "growl." It is also helpful to write these verbs on flashcards and let students physically place them in the correct column.

Once the procedure has been modeled, let students complete a set of cards, one for each column heading. Then, let them place them on the chart This will spark discussion as to the logic of their choices.

When students have finished with the class demonstrated sample, let them practice individually or in cooperative groups at their desks.

18	Identifies valid or invalid information and conclusions

Valid Pigs and Goats?

Grade Level: Middle

MATERIALS: Two stories familiar to the students

ORGANIZATION: A whole-class activity with students divided into two groups

PROCEDURE: This is a first lesson in explaining the concept of valid and invalid information. Students will have fun with this activity while learning the concept.

Begin the lesson by selecting two short (one page) stories which are familiar to the students. Some sample ideas might include: The Three Little Pigs, Little Red Riding Hood, or The Three Billy Goats Gruff. Duplicate a copy of two selected stories and cut them into pieces. The smallest sections should be in phrases and the largest sections should be a sentence or two. Place the pieces of both stories in a bag or box. This preparation should be done ahead of time.

Begin the lesson by dividing the class into two groups. For purposes of explanation, we will use The Three Little Pigs and The Three Billy Goats Gruff as our sample stories and will call the groups the "Pigs" and the "Goats." Explain to students that there are bits and pieces from two stories in the bag and it is their job to decide if a selected statement would be a part of their story. If so, then the statement would be VALID for their group. If the statement is not a part of their story, then the statement is INVALID. This is the terminology which students should use.

Beginning with a student from the "Pigs," have him come forward and select a statement from the bag. Caution the other students to refrain from calling out answers. Have the student read the statement aloud and respond with "VALID" or "INVALID." The next, and most important step, is to ask the student why it is VALID or INVALID. This reasoning is essential to the lesson.

To alleviate any pressure which students might feel about being in front of the class, you might add a feature of calling on a teammate to answer the question. Continue with this procedure, alternating between groups, until each student has had a chance to respond to a statement. There may be times when the statement is VALID for both groups. Phrases such as, "Once upon a time," or "It was a very warm day," might well apply to both stories.

18	Identifies valid or invalid information and conclusions

Validity Check

Grade Level: Middle

MATERIALS: An assortment of utensils from the kitchen, a box in which to hide them

ORGANIZATION: A whole class activity eventually dividing into cooperative groups

PROCEDURE: In this activity, students will be listing a variety of attributes of items which might be found in a given location and will then decide whether these attributes are valid or invalid for specific items.

Begin this lesson with a discussion of typical kitchen items. Have students generate a list (write it on the overhead or chalkboard) of attributes of items which might be found in the kitchen. Some sample descriptions might include: made of wood, edible, electrical, geared, round, etc.

The next step is to hold an item up for display and then call on students to decide whether each attribute listed is valid or invalid. A sample chart is shown.

	Spatula	Blender	Orange	Grater
Made of Wood	X			
Edible			X	
Electrical		X		
Geared		X		
Round		X	X	
Used for Mixing		X		
Made of Metal	X	X		

Extend this activity to smaller cooperative groups by having students generate attribute lists for items in other parts of the house -- the garage, the bathroom, the front yard, or the school room. It is important for students to use the terms VALID and INVALID rather than "yes" or "no" when they are identifying the items.

18	Identifies valid or invalid information and conclusions

Stump the Stars

Grade Level: Middle/Upper

MATERIALS: No special materials are required

ORGANIZATION: Whole class

PROCEDURE: This is a somewhat sophisticated lesson in which your class can practice "facts" or math principles in a game format.

The game proceeds as follows: one students (to be called the "Stumper") decides upon a secret function and writes it on paper which is given to the teacher. Sample functions might include: multiples of five, prime numbers, square roots, even numbers, numbers divisible by 3, etc. Other students name a random number and the "stumper" responds with VALID or INVALID. The class tries to identify the function in as few guesses as possible.

A sample game for multiples of five might be:

Classmember guesses "1"	Stumper responds "Invalid"
Classmember guesses "3"	Stumper responds "Invalid"
Classmember guesses "4"	Stumper responds "Invalid"
Classmember guesses "5"	Stumper responds "Valid"
Classmember guesses "7"	Stumper responds "Invalid"
Classmember guesses"10"	Stumper responds "Valid

Classmember guesses, "Your function is multiples of 5," which solves the problem

As an extension you can use different organizational structures for the game: 1) play with small groups with one stumper per group; 2) Use a group of children as the stumper rather than one child. This allows children who are learning a skill to collaborate on answers.

18	Identifies valid or invalid information and conclusions

Paper Machines

Grade Level: Middle/Upper

MATERIALS: Catalogues, magazines, scissors, glue mounting paper

ORGANIZATION: Individually or teams of two

PROCEDURE: Begin by having students cut out pictures of objects from magazines. It may save time to have this done as a homework assignment. Once the photos of objects have been collected, place them in a large paper bag.

First, have the students draw about ten photos from the grab-bag and hand out glue and paper. Tell them their job is to create a machine from their pictures which will accomplish some new task, or an old task in a new way. Students may need to draw connecting parts between pictures but the basic design of the machine should center around the pictures.

Some examples might be an automated toothpaste tube squeezer, automatic make-up applicator, a mobile dishwasher powered by 7-Up, etc. They are then to write up an explanation of what their machine does without identifying any of its parts. They should not share their creation at this time, secrecy is important to this activity.

When the "Rube Goldbergs" are done, collect them along with the explanations. Tape the creations to the board (without the explanations). Now have the students examine the machines and write their own ideas about what the various machines might be, based upon what they can see.

Ask for volunteers to read their ideas of other student's machines and generate some discussion as to whether or not the new explanations are valid or invalid. Questions to ponder are: Is the student wrong if his or her idea does not fit the function of the various parts? Where did the student's reasoning about the machine differ from the original? Where did it go wrong? To finish this exploration read the original explanation for each machine to judge the accuracy of the other descriptions (although this is not the goal of the exercise).

As an extension you might scramble the author's explanations beneath the machines and ask the students to identify which are correct and why.

19	Develops simple relationships among objects and draws conclusions based upon certain qualities or conditions

Birds of Paradise

Grade Level: Middle/Upper

MATERIALS: Chart of bird parts (next page), pens, art paper, crayons, Dr. Seuss books

ORGANIZATION: Individually or in groups of two

PROCEDURE: Begin by telling the students that they are going to create an imaginary bird such as Dr. Seuss would create. Students must be aware of certain facts about birds' wings, feet, and bills and their adaptive nature. Distribute the chart of birds (over) and have students review the information.

Discuss the information with students so that they see one wing type is designed for soaring, one for land-based birds that need a quick take-off, another for small birds that need to maneuver for safety, and another that glides over the ocean for hours on wind currents.

Feet too, are adapted to water, trees, catching prey in the air, walking in water or walking on land. The shape of the beak depends on the food supply that the bird finds in his environment. So the student needs to determine what the environment of his bird will be before he begins to create the bird.

After the student has drawn a bird adapted to the environment in which it lives, have him tell the class about the bird and how it uses its wings, beak and feet to survive in its environment. Birds should not be realistic except in the structure of those parts listed above. Each should be colorful and creative.

As an extension, have students write a paragraph about their feathered creations. Remember that his food supply may be imaginary as well as his environment.

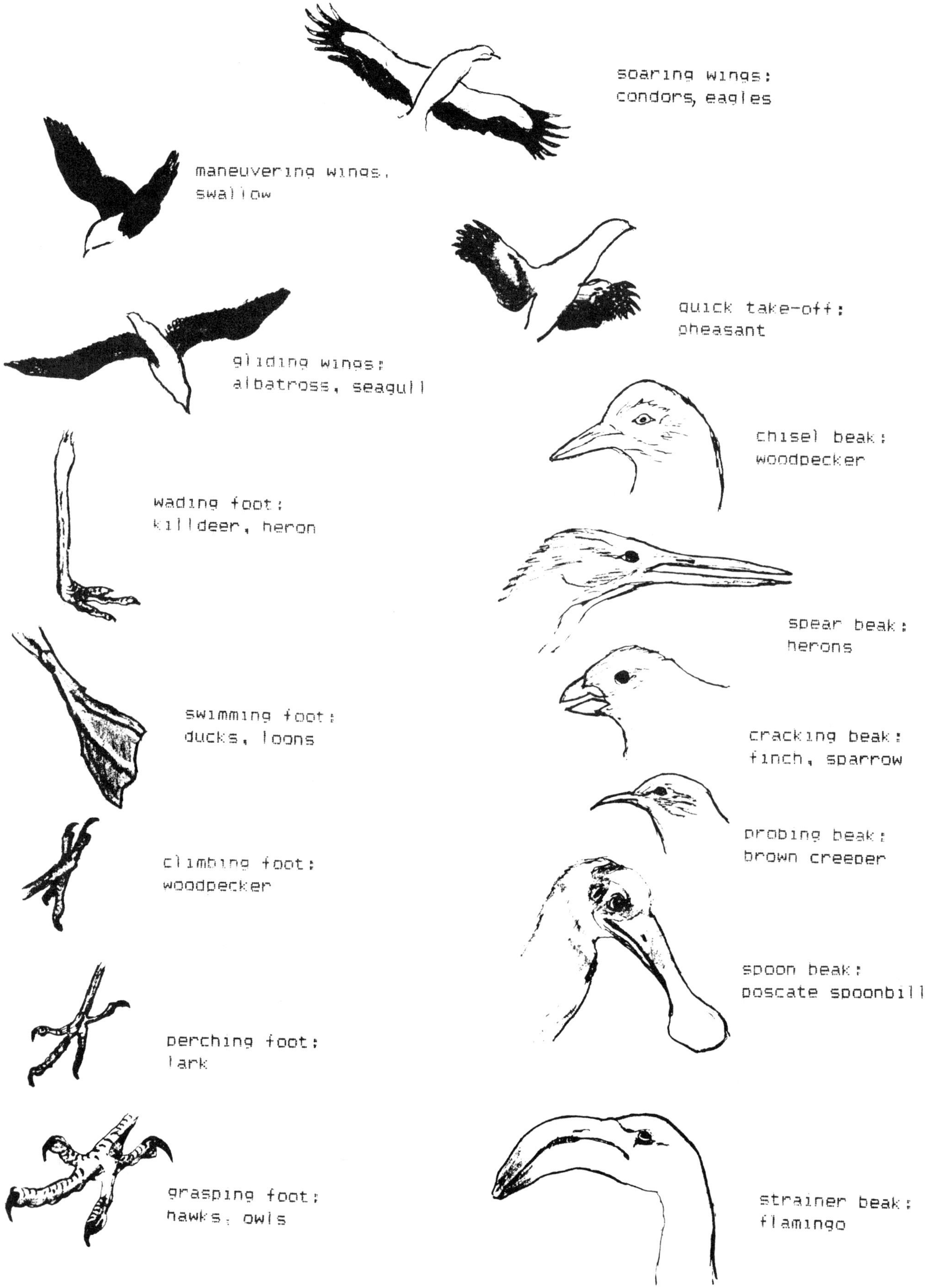
soaring wings:
condors, eagles
maneuvering wings:
swallow
quick take-off:
pheasant
gliding wings:
albatross, seagull
chisel beak:
woodpecker
wading foot:
killdeer, heron
spear beak:
herons
swimming foot:
ducks, loons
cracking beak:
finch, sparrow
probing beak:
brown creeper
climbing foot:
woodpecker
spoon beak:
roseate spoonbill
perching foot:
lark
grasping foot:
hawks, owls
strainer beak:
flamingo

19	Develops simple relationships among objects and draws conclusions based upon certain qualities or conditions

Analogy Collages

Grade Level: Middle/Upper

MATERIALS: Construction paper, markers, magazines, scissors

ORGANIZATION: Whole class activity

PROCEDURE: An important logic skill is the ability to make comparisons. This lesson will look at two different things or objects and see ways in which these object are the same. These comparisons are early attempts at analogies.

Begin by placing several samples on the board and discuss the correct answers.

1. Finger is to palm as toe is to
a) toe 2) heel 3) foot 4) sole

2. Century is to decade as dime is to
a) penny b) nickle c) quarter d) dollar

Ask students if they can think of any ways they might show these relationships in a different way. Eventually, they will come up with the idea of picture representations. Discuss how they might depict the first analogy in picture form. Repeat this procedure for sample 2 (more difficult but still possible).

At this point, hand out copies of magazines and art paper and have students cut out and create their own analogies. Some students might want to write analogies on their collages in the form of multiple choice questions, while others will simply want to present the analogy in pictorial form.

When students have finished, display their work and ask class members to solve the various analogies.

19	Develops simple relationships among objects and draws conclusions based upon certain qualities or conditions

Seeing Is Believing
Grade Level: Middle/Upper

MATERIALS: Grid paper, models of three-dimensional shapes, tape, pencils, and scissors

ORGANIZATION: Individually or groups of two

PROCEDURE: In this activity, students will attempt to draw two-dimensional figures on grid paper which would fold to match the three dimensional shapes displayed by the teacher. The object is for students to see the pattern and relationship of sides and to visualize how the folds would be made.

The teacher needs to first make one of each of the three dimensional figures to show children (if a set of shapes is available these will work as well). Explain to the children that their task is to figure out how to draw the displayed figure on the grid paper so that they can create the same three dimensional figure by folding their shape and taping it together. The children will probably make many mistakes so many pieces of grid paper will be necessary for them to complete the assignment.

It is probably best to begin with easier shapes such as a cube and rectangular prism, but eventually students should make square, rectangular, and triangular pyramids, cylinders or other polyhedrons. Real challenges include a hexagonal cylinder and a cone.

As an extension, you might want to have students create the shape by using one, continuous line (without lifting the pencil).

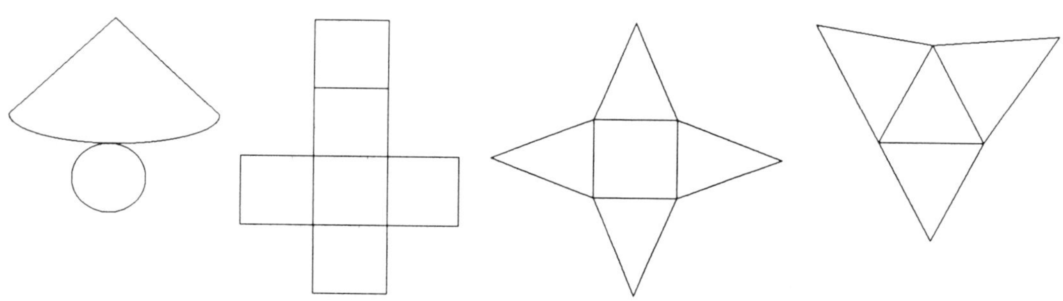

19	Develops simple relationships among objects and draws conclusions based upon certain qualities or conditions

Super Sleuth

Grade Level: Middle/Upper

MATERIALS: Clue cards, baseball diamond with spaces for names (see next page), pencils

ORGANIZATION: Groups of four

PROCEDURE: Duplicate the clue cards and the baseball diamond and cut apart. The key to this cooperative logic exercise is that the students work together to solve a problem that they might not be able to solve themselves.

Pass out the clue cards so that each student has one or more of the cards. Students work together sharing the information on their cards to solve the logic problem. Students MAY NOT give their clue card(s) to anyone else or allow anyone to read it (them), nor should they be allowed to copy down any of the information. If necessary they may use markers or other objects to manipulate in order to help conceptualize the question.

The idea is not only to solve the problem, but to do so in such a manner that the students must share their information and develop their ability to conceptualize abstract information. This procedure also helps insure that each student in the group invests some amount of interest in solving the problem.

After the students think they have solved the problem, have them read the clues again to see if their solution is reasonable according to all the information available. Students should turn in one baseball diamond picture per group with the names of players written in the correct positions which they play.

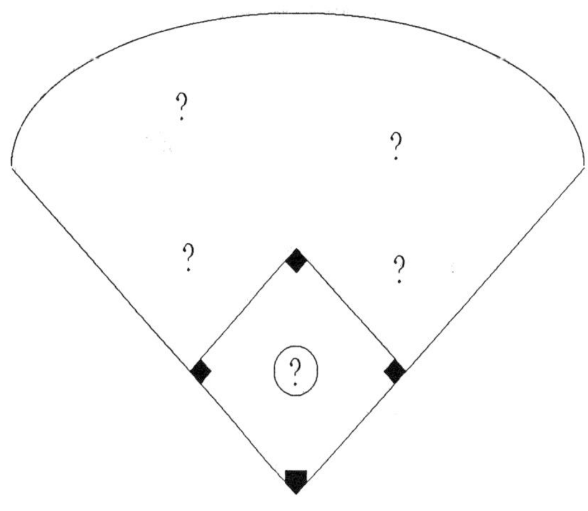

Darcy stands behind home plate	Bill is the opposite of Jose	Jorge stand between Andy and Darcy	Matt is to the left of Ron and Andy
Ron stands behind Andy	Jose is to the right of Darcy	Scott is between Bill and Andy	Ken is to the left of Andy

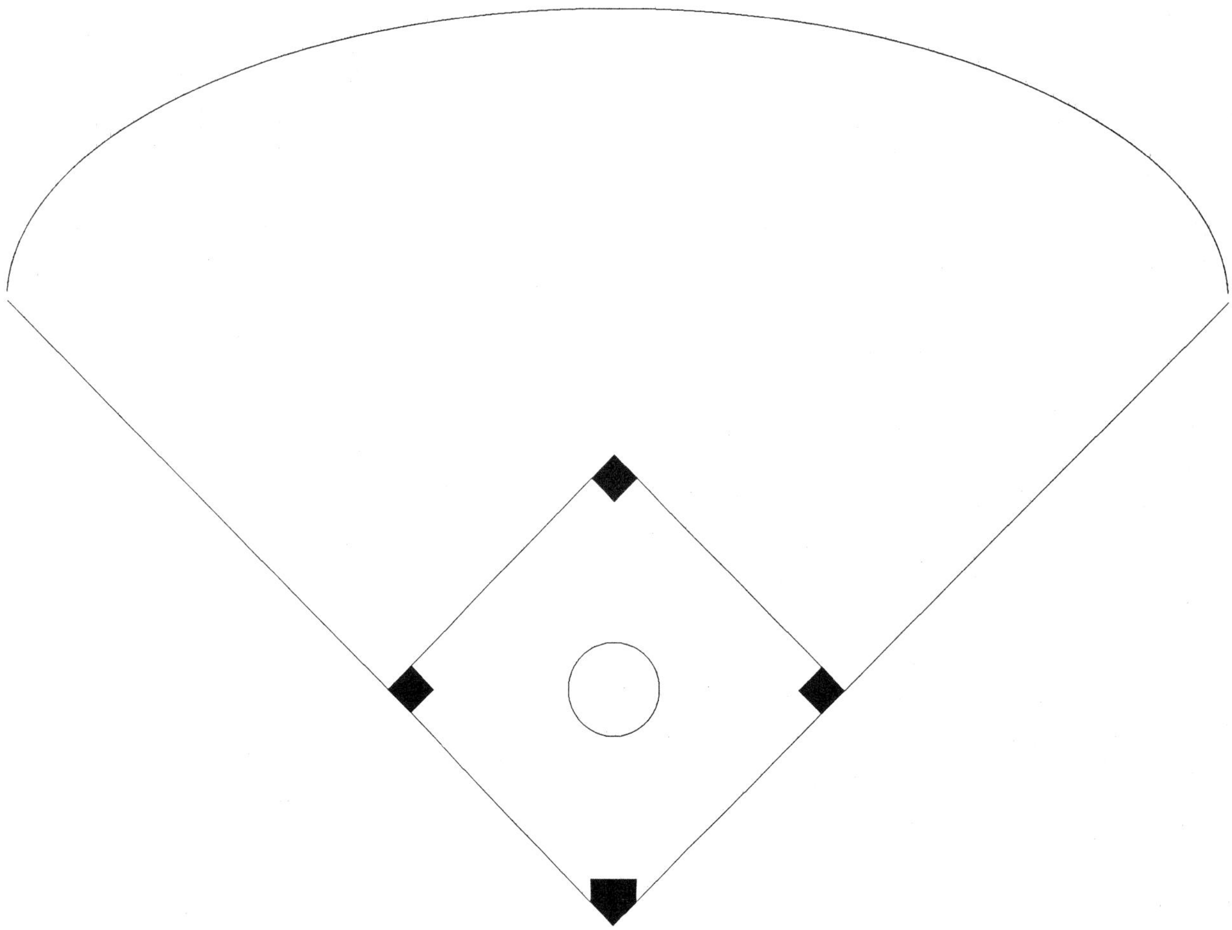

20	Solves a problem based on disqualification of available options

Your Best Guess

Grade Level: Middle/Upper

MATERIALS: No special materials are necessary

ORGANIZATION: A whole class activity then in small cooperative groups

PROCEDURE: This activity is similar to the "20 questions" game in which players ask twenty yes/no questions with the purpose of identifying a specific object. In this case, however, groups of students will be designing sets of questions which will minimize the number of questions needed to be asked to identify an item.

For example, if the item to be identified is a "diamond," students should be guided to ask questions which eliminate as many possibilities for a negative response as for a positive response. Questions such as, "Is it shiny?" or, "Is it red?" provide lots of information if true; but if false, they provide very little help in solving the problem. Questions such as, "Is or was it alive," or, "Is it natural (verses manmade)?" provide an equal amount of information regardless of a yes/no answer. This use of logic by inclusion or exclusion is the purpose of this lesson.

Begin the lesson by doing a teacher-directed game in which students receive no guidance. Keep track of the number of questions. Play a second game and discuss the value of each question asked by students. Students who use inappropriate questions should be guided to change them into more effective questions.

Have a student be the "teacher" for the third game and continue to guide students' questions. The questions below show a series of effective (and logical) queries for identifying the diamond.

Is it man-made? NO Therefore, it is natural
Is or has it been alive? NO Therefore, it must be mineral
Is it valuable? YES Must be a precious gem
Can light shine through it? Yes
Is it a diamond? Yes

20	Solves a problem based on disqualification of available options

No Objections

Grade Level: Middle

MATERIALS: An assortment of toys, a variety of different colored shapes, or items available in the room

ORGANIZATION: Groups of two, three, or four

PROCEDURE: This is a lesson using riddles to eliminate possible answers to a given question. Begin the lesson by selecting an item in the room and say, "I am thinking of an object in the room."

"It is not red, it is not square, it has no wheels, it is long and thin, what is it?"
Add more statements if the first few guesses are incorrect.

Once students understand, divide the class into small groups and have them choose a leader to play the same role the teacher did in the model game. If you are using toys or shapes, spread them out across the table. There should be six to twelve shapes (samples shown below) in a variety of colors. Sample statements for shapes might include, "It is not small, it is blue, it has no curves, which is it?"

Allow each student in the group to be a leader for several games.

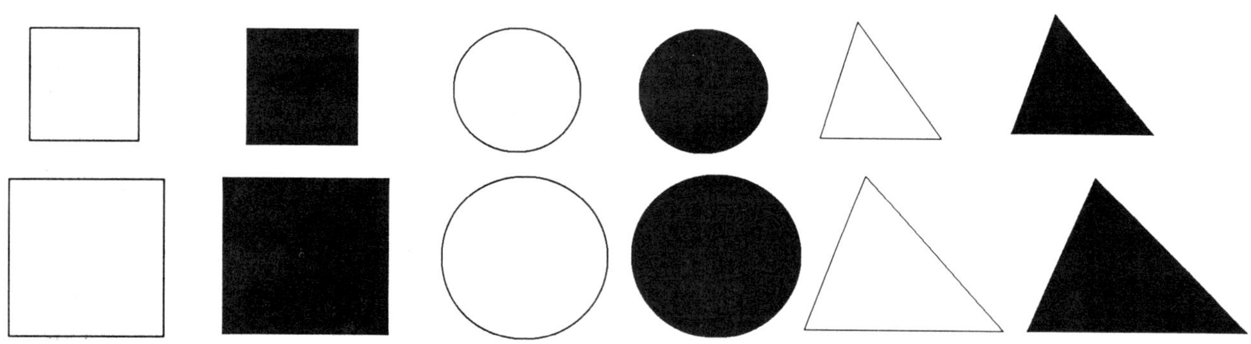

20	Solves a problem based on disqualification of available options

Powder Puff Attributes

Grade Level: Middle/Upper

MATERIALS: Assortment of white powdery materials (flour, powdered sugar, cornstarch, baking powder, baking soda, baby powder, bath powder), charts for recording

ORGANIZATION: Groups of two, three, or four

PROCEDURE: This lesson is based on the old "mystery powder" lesson used by science teachers. Each powder should be in a small numbered container. You do not need a large amount of each powder.

The students should guess what each mystery powder is by feeling, touching, smelling, observing, or tasting (warn the students that they should never taste or smell an unknown substance, but in this case, the teacher knows that the substances are harmless). You should warn them not to taste the non-edible items, (e.g. "do not taste samples 2 and 5). They should create a list for each powder which identifies attributes and explanations why they have made their decisions.

After guessing and discussing what each individual powder is, this lesson can be extended by mixing the powders and having students identify the powder which is not in the mix.

20	Solves a problem based on disqualification of available options

Color, Tac, Toe

Grade Level: Middle

MATERIALS: Grid, colored markers, two colors of chalk

ORGANIZATION: Teams of two

PROCEDURE: The object of this activity is to have students develop a logical strategy that will allow them to get three squares of their color in a row either across, down, or diagonally.

Begin the lesson by demonstrating a game on the chalkboard. Draw a 3 by 3 grid and select one student to come forward and be your opponent. You will use one color of chalk, your student a second color. Your opponent may begin by coloring one or two squares of his/her choice, the squares need not be adjacent, merely uncolored. The object of the game is to get three in a row. Play the game allowing students to ask questions as moves are made. Two sample games are shown.

GAME 1

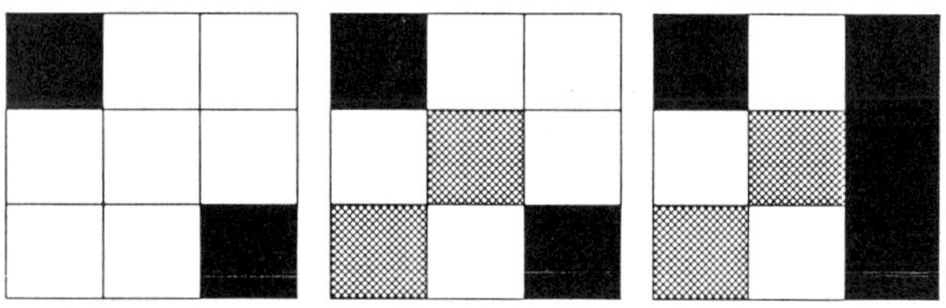

First Turn Second Turn Third Turn

GAME 2

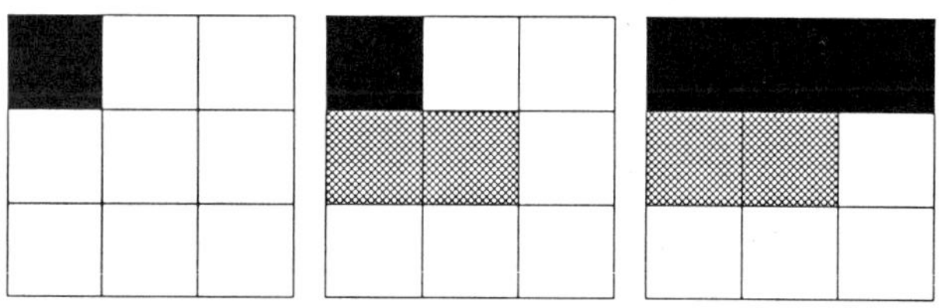

First Turn Second Turn Third Turn

Give students sheets of grid paper and divide them into teams of two. Let them play the game. After fifteen minutes or so, stop the play and have the students discuss any strategies they may have developed. After a few ideas have been exchanged, explain that there may be some strategies for choosing squares to color, as well as overall strategies for winning. The concept here is that sometimes the logical options available are determined not by an individual's rational choice but by circumstances which limit the range of possibilities.

21	Demonstrates an understanding in using the words: AND, OR, and NOT

To And or Not to And

Grade Level: Middle/Upper

MATERIALS: Pocket charts (if available) or tape, cards, sentence strip cards, marking pens

ORGANIZATION: A whole class activity then dividing into teams of two, three, or four students

PROCEDURE: In this activity, students will experiment with the logic of adding AND, OR, or NOT to various sentences.

Prepare sentence strip cards ahead of time a sample is:

Jim		John	bought food
They	are		hungry

Make three more cards with a key word (AND, OR, NOT):

AND	OR	NOT

Have several other blank cards available to write other words, since the key words will necessitate grammatical changes.

Call on a student to come forward and place a key word in one of the blank spaces. Ask students to decide if the added key word is logical. For example, if the student placed NOT between Jim and John, does it make sense that both are hungry? Students may decide that THEY will need to be replaced by HE.

Continue this procedure until students have had time to experiment with many different combinations. Then, have the class form teams of two, three or four students and have each group generate a set of cards similar to those shown and experiment with the logic of AND, OR, NOT.

21	Demonstrates an understanding in using the words: AND, OR, and NOT

Bits and Pieces

Grade Level: Middle/Upper

MATERIALS: Various lengths of colorful yarns (2" to 9") for each group, rulers, tape, and art paper

ORGANIZATION: Cooperative groups of three or four students

PROCEDURE: This is an early lesson in understanding the use of the words AND, OR, and NOT in which students glue various pieces of yarn to demonstrate statements which they have created.

Give each group a set of eight to twelve random lengths of yarn and have them select three pieces. They should place them in random order and label them A, B, and C. It will be more effective if students label their strings differently. This will cause each statement they make to be different and will enhance the "logical" thought in checking one another's responses. Ask each student to make up a statement about the three pieces using the words AND, OR, or NOT. Sample statements might include:

A is longer than B or C; B and C are not longer than A; A or B are longer than C

Have each group select a statement (or statements) and write it (them) on a piece of art paper, gluing the three pieces of yarn and labeling them A, B, or C. The teacher may want to create a sample (shown) to clarify this procedure (see illustration).

Once students have completed this procedure hold up two or three samples for the class to discuss and evaluate.

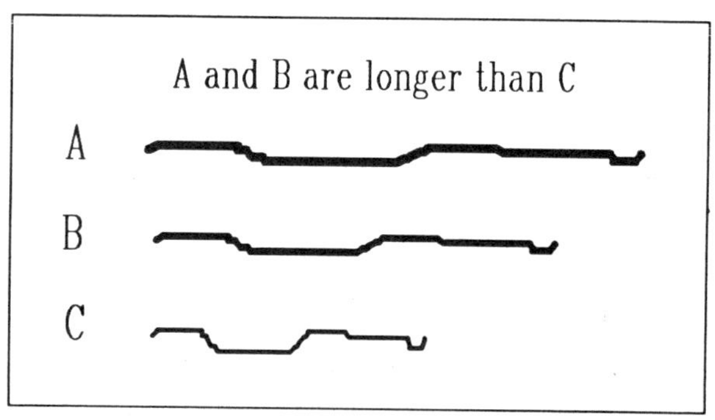

For students who have mastered this concept, you might have them extend the lesson to include statements about four or more pieces of string. Students should create a representation for the statements they create.

21	Demonstrates an understanding in using the words: AND, OR, and NOT

Selling Soap

Grade Level: Middle/Upper

MATERIALS: Soap for baths, clothes, walls, floors, cars, silk, wool, hair, dogs, etc. (containers need not be full)

ORGANIZATION: Individually or in groups of two or three

PROCEDURE: Explain that in the field of logic the use of a certain vocabulary allows students to make statements which are clear and which explain the relationships among various bits of information. Using the proper language will help students explain the similarities and differences more accurately.

Begin by saying that soap is a commodity that cleans, but many soap products have only one type of application while there are a few that have multiple uses. The task here is to determine the use or uses of each of the soap samples and then classify them by using a vocabulary that will clearly delineate the usages. Students should use the soap containers to verify the uses.

For example: Dish soap can be used for dishes AND/OR laundry
Dog soap can be used for dogs AND/OR cats, but NOT for humans
Soap for walls and floors can NOT be used for personal hygiene

An ideal structure for this lesson is to divide students into teams of two and to provide (or have students provide) a different soap product for each team. In this way, students can write a statement about a soap type and then "pass" the various products around the room. This will give each team the experience of writing 12 to 16 (depending on class size) AND, OR, NOT statements.

When students have completed the research, see if they can sort the soap types by attributes, i.e. all hygiene soaps in one category, all non-human consumption soaps, etc. This lesson could also extend into a very sophisticated Venn diagram with the entire class working together to complete this project.

21	Demonstrates an understanding in using the words: AND, OR, and NOT

Nailed to the Wall

Grade Level: Middle/Upper

MATERIALS: A wide assortment of nails (can be purchased at a lumber yard), containers, paper and pencils

ORGANIZATION: Groups of three or four students

PROCEDURE: In this activity, students will be analyzing a variety of nail types to decide upon their uses and writing AND, OR, NOT statements to describe these uses.

Begin by handing out a container of ten to twelve various nails to each group and give students an opportunity to speculate about their uses and why they have certain attributes (color, coatings, thick shanks, etc.). They should experiment with grouping the nails in a variety of configurations.. Allow sufficient time for students to become familiar with the variety of the nails and possible uses. Then have each group write statements which would enable classmates to select certain nails from the group. The goal is for students to match statements with proper nails,

Some sample statements should include:

> Two of the nails have large heads and are probably used for holding soft materials.
> Three of the nails are not used for thick boards.
> Two of the nails are designed for easy removal or for temporary jobs.
> Half of the nails are not specially designed.

Have students write these statements on a piece of paper then let groups exchange their containers of nails and statements to see if the new group can properly match them.

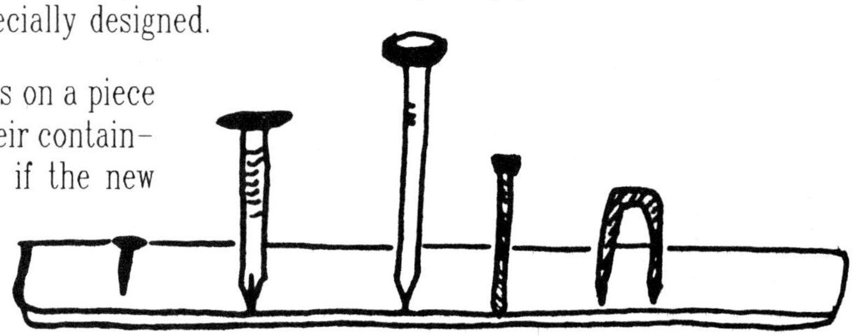

22	Interpolates information into a set of data

Interpol Sleuth

Grade Level: Middle/Upper

MATERIALS: No special materials are necessary (access to almanacs and atlases might be helpful)

ORGANIZATION: Groups of three or four

PROCEDURE: In this lesson, students will interpolate data by creating scenarios involving kidnappings, bank robberies, etc. and will create clues which they will use to solve cases.

Begin the lesson by explaining the following conditions: a bank was robbed in a European city and that same night several marked bills were exchanged in New York which had been traced to the robbery money. As a member of interpol, your only clues are: 1) the person who exchanged money mentioned the word "Bobby;" 2) he had an accent but spoke English; 3) he spoke to one other person and said something about a tower. Where would you begin your investigation and why?

This type of an activity provides an open ended solution. The obvious response is London but students may use logic to create other solutions as well. The emphasis should be on "logical conclusions" rather than "right answers."

Divide students into groups of two, three, or four students and give them time to develop their own case scenarios. Then allow them to describe the clues to one another to see if they can "solve" the case. Students need not have a specific location in mind; by making up clues they can actually have fun putting random clues together to discover possible locations.

There is an obvious relation to geography in this lesson. Encourage students to use various resources to create and to solve these logic stories.

22	Interpolates information into a set of data

Comic Art

Grade Level: Middle/Upper

MATERIALS: Enough comic strips for each team or each student, scissors, glue, art paper, crayons or felt pens, rulers

ORGANIZATION: Teams of two or individually

PROCEDURE: Prior to the lesson, cut out and discard the midddle frame or frames of the comic strips and glue the beginning and end frames on art paper leaving a blank space for missing frames between the beginning and end frames.

Hand out papers to the students or let them choose which one they want to work on.

Tell the students that they will draw scenes and write in conversation bubbles in the blank spaces. The scenes should fit logically between the printed ones. Use a ruler to make neat frames. After drawing with a pencil they can use crayons or markers to fill in the color.

When all are done, either let each student or team read their strips aloud or exchange them with their classmates. Discuss and critique the continuity of the frames they created.

22	Interpolates information into a set of data

Lost and Found Findings

Grade Level: Middle/Upper

MATERIALS: Paper, pencil, tally sheets, lost and found department

ORGANIZATION: Teams of two or three

PROCEDURE: Explain to the class that they will be examining the items left in the lost and found area to interpolate certain information. Ask what information they might be able to infer from examining the lost articles. Answers might include: the type of articles lost most frequently, by whom these articles are lost (boys or girls, younger or older children), what colors, what sizes, or what styles are lost most often.

Allow students to examine the items left in the lost and found and to tally their findings. They should then project reasons for the items ending up in lost and found. Factors which might be considered include:

1. Temperature and appropriateness to weather
2. Age of person who lost the item (determined by size or attractiveness of the item)
3. Attractiveness or desireability of the item (Is it the "in" thing?)

As an extension the teacher may plant some uncommon items (work boots, hard hat, apron) and ask students to identify the erroneous item and challenge them to identify the bogus planted items (a dress shoe, dress glove, etc.).

22	Interpolates information into a set of data

Oh To Be a Wall Flower!

Grade Level: Middle

MATERIALS: Wallpaper books, drawing paper, markers, rulers

ORGANIZATION: Individually or in teams of two

PROCEDURE: Students will probably not be aware that wallpaper books are displayed in a particular sequence. They will be discovering these patterns and then drawing wallpaper samples which are missing from pattern sets.

Begin by displaying a wallpaper book's first page and ask students to describe the paper using attributes. Turn to the second page and ask the class what attributes have changed from page one. Continue with this process through several pages. You may want to repeat this process with a second book.

The student activity part of the lesson is best organized by the teacher cutting squares of wallpaper samples (6 to 12 pages in 6" by 6" squares) and then substituting blank pages (this is the process of interpolation) of art paper for one or two of the wallpaper samples. Staple the pages together and distribute to students.

Their task is to draw the pattern, color, and design for the blank page so that it keeps the correct sequence. After they have interpolated these sequences, let students compare their drawings to the actual wallpaper sample. These can be displayed or used for discussion as a class. There are many different answers possible -- all of them correct, as long as students have a logical justification for their drawings.

23	Extrapolates information from a set of data

The Great Chase

Grade Level: Middle/Upper

MATERIALS: No special materials are necessary although Almanacs, Atlases, maps and social studies books are helpful

ORGANIZATION: Groups of two, three, or four students

PROCEDURE: This activity is related to the Interpol lesson of task analysis 22 but it extends the interpolation of using existing clues to an extrapolation problem of finding a pattern of future actions.

The concept of the lesson is to ask students to predict locations of future crimes based on the past crime locations.

The teacher should begin the lesson with explaining this scenario: A group of bank robbers has been terrorizing the United States. They have staged five robberies in five different cities. Their methods are always the same but police are never certain where they will strike next. To date, they have robbed banks in Albuquerque, Denver, Grand Rapids, Jacksonville, and Minneapolis. The police should anticipate the location of the next robbery based on this information.

Elicit this pattern from the students: the robberies occured in cities which begin with every third letter of the alphabet. You might create a second pattern which is more difficult. Patterns can consist of letter combinations, sounds, direction, travel distance, etc.

Divide students into groups and let them create their own patterns. Once again, the purpose of this activity is not to find the "right answer" but to find a list of "logical deductions" based upon logical thought.

Once students have written their patterns have them exchange information within groups to solve the puzzles.

23	Extrapolates information from a set of data

Trashology

Grade Level: Middle/Upper

MATERIALS: A collection of trash from the waste basket from another classroom for each group

ORGANIZATION: Groups of four students

PROCEDURE: Tell the students that by using logic they can find out what has happened in a given location in the past. By analyzing the results (or byproducts) of an activity they can tell what activities actually transpired.

Give each group a trash container as described in the materials. This container needs trash from an active classroom that can give students a good sample of work that took place in the classroom. The teacher will need to preview the collection to be sure "undesirable" items are not found. You may wish to make an arrangement with a fellow teacher.

Have students remove items, keeping them in a semblance of order. They will need to decide upon a method of organizing their "data." Once they have finished organizing their "educational effluvium," have them write out conclusions about what went on in that classroom (what learning and what grade level).

After students have written their conclusions, have each group make a poster of their findings, linking it to the pertinent clues. Have them spread their "clues" out on their tables so that classmates can circulate the room and inspect each other's work. Allow them to challenge the findings of another group (this could generate a good logic discussion). In order to check their work, you may also let them present their conclusions to the class from which the evidence was taken to check for accuracy.

An extension might be to "salt the lode" with a "bogus bag" of false trash and see if the students can identify which one is the plant.

23	Extrapolates information from a set of data

For the Record...

Grade Level: Upper

MATERIALS: Pencil, paper, graph paper, world almanacs for the past several years, encylopedias, or The Book of World Records

ORGANIZATION: Individually or teams of two students

PROCEDURE: Explain to the class that extrapolation of information means that they will be taking information from a set of data and using that information to predict future occurrences.

Tell them that they will be looking up sports records in world almanacs and encyclopedias and will be predicting the sports records of the future, what records will be broken and by how much. Each student might take a different event to research and to prepare a report for the class. Swimming and track and field sports are the most effective to use for this project.

Begin by having students graph the results of past records to help predict future records. They should also make predictions regarding the point at which records will no longer be broken; in other words, how high (low) can the records reasonably be expected to go given the limit of human potential. They may also want to predict what the record will be in the year 2000.

POLE VAULT RECORDS

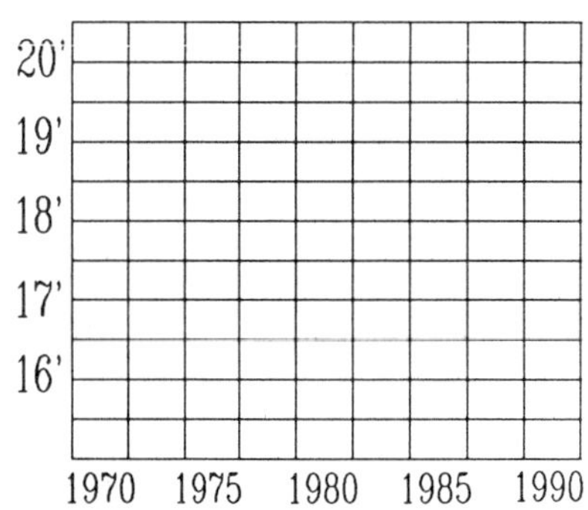

As an extension, you may wish to have students do a research project to find out why a long standing record was suddenly broken by several people in a short period of time. Were there developments of new equipment, or was the old record a psychological barrier which intimidated performers? This can lead to some very interesting class discussion.

23	Extrapolates information from a set of data

And Then...

Grade Level: Middle/Upper

MATERIALS: A comic strip for each student, scissors, glue, paper, pencils, crayons or felt pens, rulers

ORGANIZATION: Individuals or teams of two

PROCEDURE: Prior to the lesson cut off and discard the end frames of the comic strips. Pass out the comic strips to the students. You can assign strips or let the students choose the strips on which they wish to work.

Have the students glue their strip on paper leaving space on the right side. Use a ruler to draw neat frames from the end of the strip to the right edge of the paper (creating final frames).

Tell the students that they are to create appropriate pictures and dialogue for the ending frames so that the comic will have a logical conclusion. When all are done, display the finished strips. Discuss and critique the appropriateness and logic of the created conclusions.

23	Extrapolates information from a set of data

Who Views What?

Grade Level: Middle/Upper

MATERIALS: A television and T.V. Guide.

ORGANIZATION: Groups of four

PROCEDURE: In this activity, students will do research on television programming and advertisements and try to make predictions about why networks schedule them as they do.

Begin by explaining that television networks spend millions of dollars on marketing studies, the time programs are on, and the "line-up" of shows for the evening. Advertisers also do this same type of marketing study to decide upon the best times to advertise their products. Tell students that they will be forming groups to do research on television programming and advertising.

Have each group meet to decide upon a time period and a range of television stations to study. One group might take Saturday morning from 8:00 to 10:00, another group the 4:00 to 5:00 time slot on Tuesday afternoon. Each group member should take a different television station to monitor.

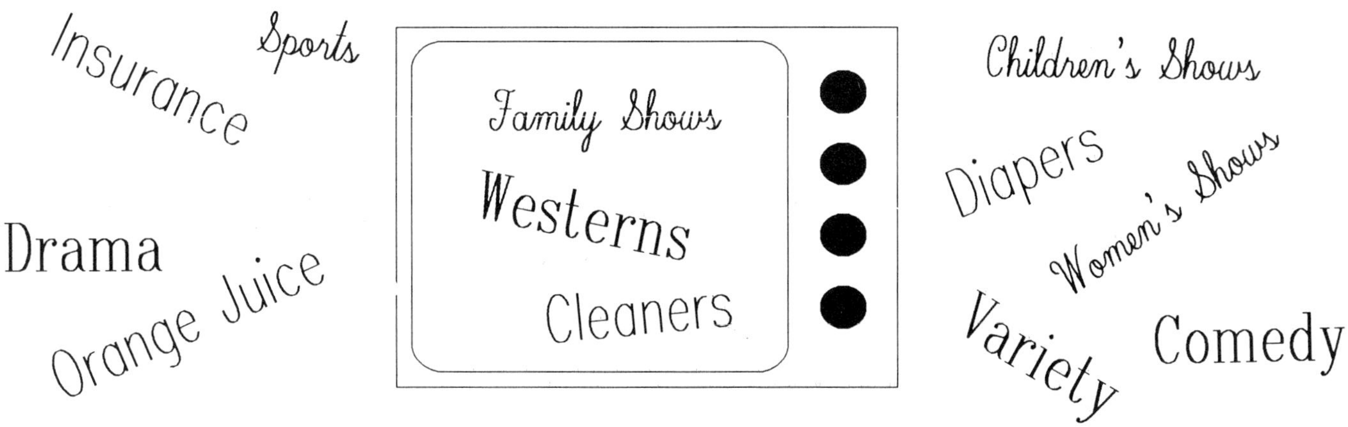

Students should take notes on the programming, such as the types of characters in the programs, the types of viewers (age, sex, etc.), the type of program (comedy, drama, entertainment), the types of commercials presented during breaks, and other information which students feel might be important.

Once data has been collected, allow groups time to extrapolate the information into hypotheses. For example, one group might state, "CBS and ABC have half hour comedies from 8:00 to 9:00, and NBC had a special for children. The special was probably shown on this night and at this time because children would not want to watch the comedies, and it was an early in the evening before bed time. The commercials on ABC were for toys and diapers, so there were commercials for adults (children don't buy diapers) as well as children."

Once each group has determined the "logic" of the marketers, have them present their findings and postulations to the class.

24	Identifies relationships among objects according to position in a pattern or sequence

Shapes of Things to Come

Grade Level: Middle/Upper

MATERIALS: Colored shapes assortment for each group. One sheet of paper on which to lay separate groups of shapes. Circles of string may be used instead of paper

ORGANIZATION: Groups of two

PROCEDURE: The students will identify the relationships among the shapes according to a prearranged pattern.

The teacher should model the activity by laying out a pattern of shapes such as those described below. Ask the students to sort the items into logical groups and explain the ways the items are the same or different.

Sample set 1: 3 red shapes, 3 green shapes, and 3 blue shapes.
Sample set 2: 2 circles, 2 squares, and 2 triangles (small and large)

Have the students rearrange the sample set to see if they can group the shapes based on a different attribute. Each time they should explain the logic of the grouping. Once students understand the concept, have them divide into teams of two and give them a set of color shapes. Let them arrange them and explain their groupings to one another.

You may want to extend the lesson by having students cut out their own shape sets to divide by attribute.

Sample Set 1

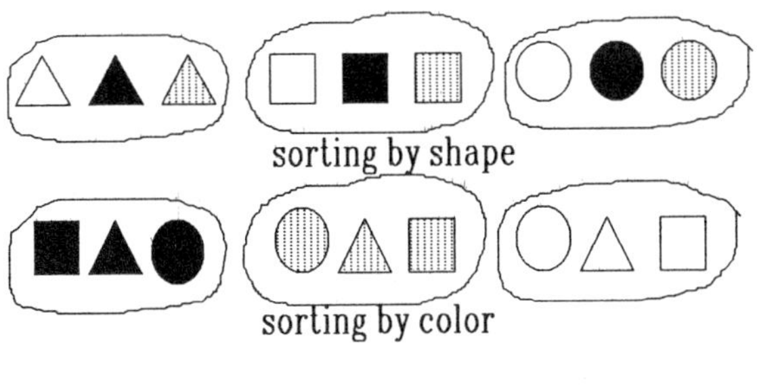

Sample Set 2

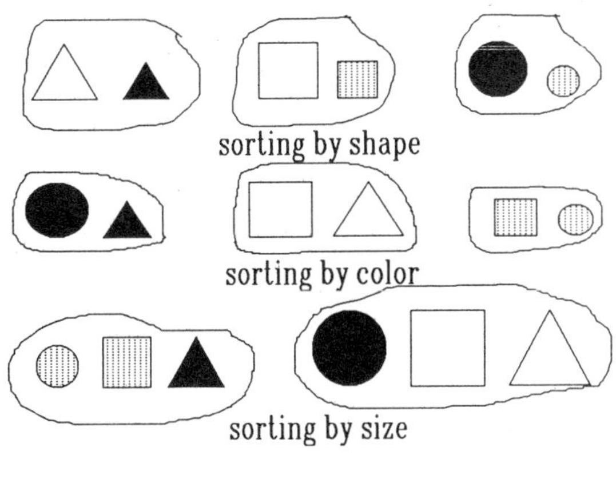

24	Identifies relationships among objects according to position in a pattern or sequence

Puzzling Pictures

Grade Level: Middle/Upper

MATERIALS: Pictures (simple line drawings, see below), tracing paper, scissors, glue, pencils.

ORGANIZATION: Groups of two

PROCEDURE: Students will be identifying relationships among objects according to their position in a pattern and selecting the picture that does not belong in the pattern.

Prior to the lesson, the teacher should duplicate the examples given below. Using these examples, have students identify the picture that is different from the others, have them state the difference, and explain how they identified this difference.

Once students understand the concept, explain that they are to make up their own sets of pictures. For younger students, the teacher can trace these from other sources or duplicate them from cartoons or line drawing books. Older students can trace their own or create their own drawings.

Have students glue their picture sets on a sheet of paper. Always make sure that one is different in some minor way. If necessary provide teacher-made sets for extra practice before the students make up their own.

Have students exchange sets and see if they can solve one another's "logic puzzles."

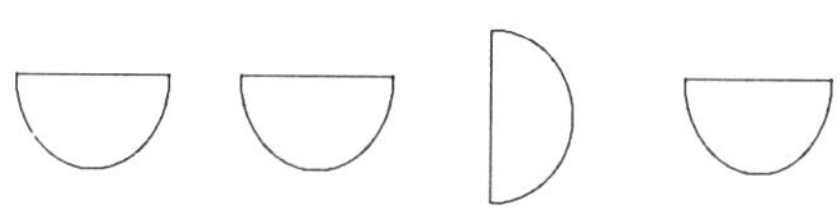

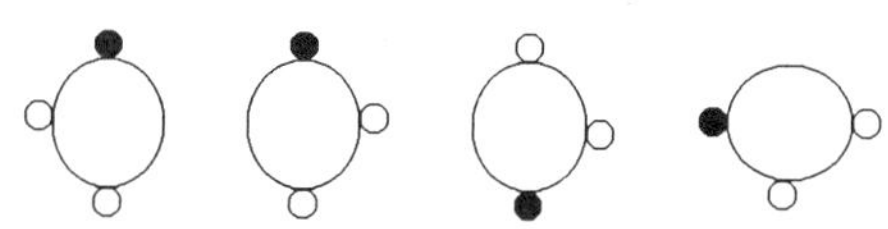

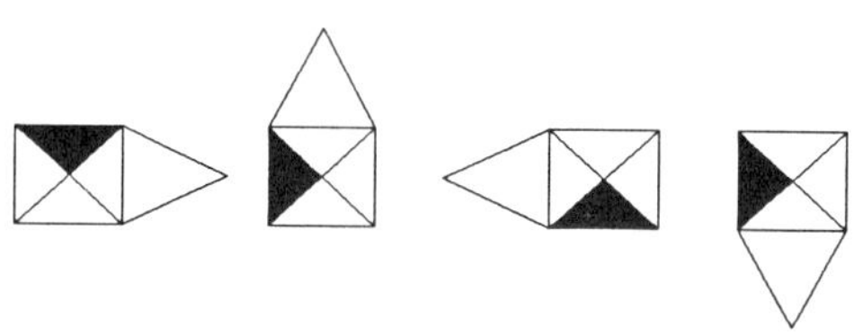

24	Identifies relationships among objects according to position in a pattern or sequence

Up, Down, Everywhere

Grade Level: Middle/Upper

MATERIALS: Several sheets of wallpaper, gift wrap, or patterned paper.

ORGANIZATION: Groups of two, three, or four

PROCEDURE: Begin by brainstorming with students for attributes that might be used to create a pattern or sequence (attributes which can be used to identify the basis of creating a pattern). Record this list of words on the board. This list should include words such as: position, size, shape, color, or design details.

Explain to the students that they will be finding these attributes in wallpaper samples and will write the attributes which create a pattern on index cards as clues for their classmates. The teacher may want to model a sample for the students:

"The pattern on my wallpaper is created by design details, size, and position."

Can you tell which wallpaper I am describing? Hold up two or three samples.

At this point, if students understand the concept, hand each student (or team) a wallpaper sample and an index card and have them complete a statement similar to the the teacher model. Once all groups have completed this activity, have students place wallpaper samples around the room. Select individual students to try to match the index card to the wallpaper samples.

24	Identifies relationships among objects according to position in a pattern or sequence

Noodle It Through

Grade Level: Middle/Upper

MATERIALS: Macaroni noodles: small elbow, large elbow, spaghetti, rotini (or any combination of four types of macaroni)

ORGANIZATION: Individually, pairs, or teams

PROCEDURE: Explain that one of the skills involved in logical thinking is the ability to link things together in proper sequence. In electronics the various components -- resistors, capacitors, silicon chips, etc. -- must all be linked together in sequence before a system will work properly.

Tell the students that they will be building various circuits out of noodles (was it Marconi or macaroni who invented the radio?). Each noodle, however, has a special quality.

Small Elbows: allow electricity to flow only to the right
Large Elbows: allow electricity to flow only to the left
Spaghetti: allow electricity to flow only horizontally
Rotini: allow electricity to flow only vertically

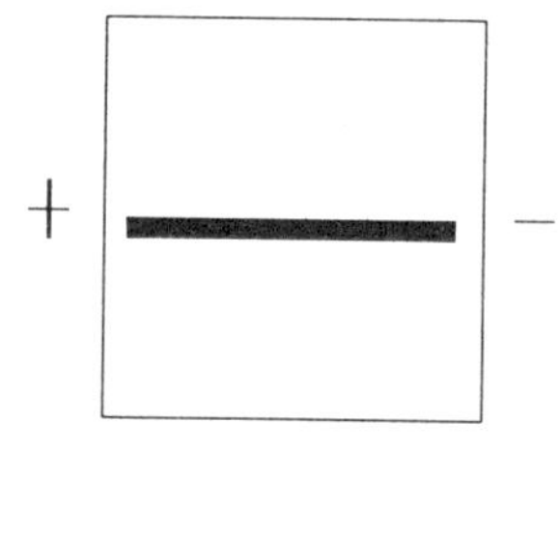

Provide students with a simple example on the board. Draw a large square with a positive sign on the left-hand side and a negative sign on the right-hand side. Now ask for a volunteer to channel the electricity from the positive side to the negative. Point out that electricity always flows from positive to negative. They should be encouraged to start by using as few noodles as possible (altough they may need to adjust the length of the noodles). After the students have demonstrated that they are familiar with the concept, add another square and place positive and negative on various sides of the square (or board) and see if they can solve this noodle puzzle.

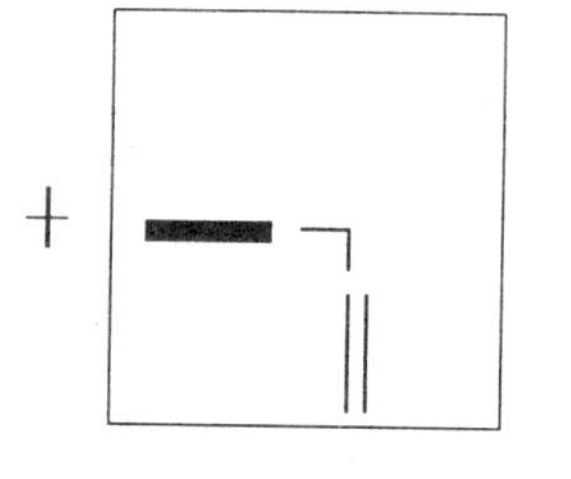

There are many different variations possible at this point. You may want to provide copies of "circuits" to for your class to "noodle through." Or you may wish to let your students create their own circuits with a specified number of "appliances" (squares with a plus and minus sign). Students should also be allowed to build their own blank circuits with which to challenge other students (these can turn into timed events). Ultimately, students can do a wiring diagram of their house (electricity enters on one side, flows through all the appliances on all floors and rooms, and out the other side). Many extensions are possible by adding new noodle pieces, obstacles, items to change direction, etc.

25	Explains logical relationships among objects by writing accurate syllogisms

Mix and Match Syllogisms

Grade Level: Middle/Upper

MATERIALS: Sentence strips cards and markers or other materials for display of written syllogisms

ORGANIZATION: Groups of two

PROCEDURE: In this lesson students will be learning about syllogisms by making deductive statements based on implied (general) information and working them into syllogisms which include the implied information.

The teacher should begin by describing the following scenario. Suppose two friends are sitting at home and it is raining hard outside. One says to the other, "If it rains like this all afternoon the baseball game will be cancelled." The friend has just made a DEDUCTIVE STATEMENT. If the other friends asks, "How do you know?, the first person must explain the logic by saying that people don't play baseball when it is raining hard.

The first friend has given a REASON for a GENERAL STATEMENT which they have accepted as true. When you make a deduction you do not always explain the general statement upon which the reasoning is based. In every case, however, there must be one accepted general statement by which you arrive at a conclusion or make a DEDUCTION. Look at the following statements and match the deduction with the general statement that is the basis for your deduction.

If this lesson seems incomplete
And if incomplete lessons are continued
Then this lesson is continued

1. When 379 is multiplied by 4, the product is an even number.
2. If Jim attends Midtown Academy next year, he will take part in some sport.
3. If Ann was home last night, she was not at the basketball game in the school gym.
4. If you are to be an engineer, you should study mathematics.
5. If a steel wire is heated, it will become longer.

a. Many students at Midtown Academy are good athletes.
b. All right angles are equal.
c. Steel expands when hot.
d. All students at Midtown Academy <u>must</u> take part in some sport.
e. Girls do not attend basketball games.
f. The study of engineering requires a good foundation in math.
g. The product of any integer and an even integer is an even number.
h. A person cannot be in two places at the same time.

Write these statements as syllogisms using the pattern of: IF...AND....THEN. Have students work together to create four statements of deduction, separating the accepted general statement as above. Exchange papers and have each group of two work to solve the new list and write in a syllogistic statement.

Some sample syllogisms would include:

If engineers need a foundation in mathematics,
And if you want to be an engineer,
Then you should study mathematics.

If Ann was at home all last night,
And if the basketball game was not at Ann's,
Then Ann was not at the game.

25	Explains logical relationships among objects by writing accurate syllogisms

Not So Silly-gisms

Grade Level: Middle/Upper

MATERIALS: Word cards, markers

ORGANIZATION: A whole class activity then dividing to groups of two, three, or four

PROCEDURE: This is a beginning lesson in working with syllogisms.

Begin by elicting a series of related items to write on the chalkboard or overhead. Some samples might include: Elm, tree, leaves, green; or Hershey, candy, sweet, wrapper. Using one of these sets, or a student generated set, write each word on index cards so you have two sets (see below).

Elm	Tree	Leaves	Green
Elm	Tree	Leaves	Green

Write the following structure on the chalkboard or overhead:

IF __________ __________
AND __________ __________
THEN __________ __________

The rules in placing cards in the blanks are 1) only three of the four words can be used, 2) the same two words cannot repeat on any line, 3) each word must be used two times.

An example: IF Elms are trees
AND trees have leaves
THEN Elms have leaves

Try rearranging cards using different verbs to see if students can create numerous syllogisms. At this point, it is not necessary to analyze for valid and nonvalid syllogisms but you may want to ask students to decide whether or not the syllogisms they have created are true.

As a final step, divide students into teams of two, three, or four and give them index cards so they can experiment with their own syllogism creation and structure.

25	Explains logical relationships among objects by writing accurate syllogisms

Human Syllogisms

Grade Level: Middle/Upper

MATERIALS: Word cards, markers, string (optional) with card prepared ahead of time

ORGANIZATION: A whole class activity with students forming their own groups

PROCEDURE: Begin by preparing two set of cards for your class. The words to be written on each card are shown in the box (bottom of page). You will notice that there are several duplicates and related words. In this lesson, the students will each be given one of the cards and it will be their task to find fellow students to stand with in order to form a syllogism.

Begin by having six students come forward. Give each student a word card with one of the following words written on it: peaches, peaches, pits, pits, seeds, seeds. Tell the student that it is their task to organize themselves into a human syllogism using their word. They must stand in the correct order and they will have to provide the extra wording leading into their word. For example,

Student 1	"IF PEACHES"
Student 2	"HAVE PITS"
Student 3	"AND IF PITS"
Student 4	"ARE SEEDS"
Student 5	"THEN PEACHES
Student 6	"HAVE SEEDS"

When the class understands the concept, hand each student a word card from set one and have students roam the classroom looking for his unknown syllogism. Once each student has found a group, they should practice "performing" their syllogism for presentation to the class. Repeat this procedure with set two.

Card Set 1		Card Set 2	
light bulbs (6 cards)	plugged in (2 cards)	cowboys (2 cards)	saddles (4 cards)
electrical energy (6 cards)	batteries (4 cards)	horses (4 cards)	leather (2 cards)
flashlights (4 cards)	some toys (2 cards)	hay (4 cards)	cowhide (2 cards)
switches (4 cards)	noise (2 cards)	food (2 cards)	milk (2 cards)
		crops (2 cards)	cows (2 cards)
		farmers (4 cards)	

26	Uses Venn Diagrams to represent complex relationships (sets, subsets, etc.)

Sneaky Venns

Grade Level: Upper

MATERIALS: Collection of magazines, glue, scissors, poster/mounting paper

ORGANIZATION: Indivdual or groups of two

PROCEDURE: Logic involves several skills which call for the observer to draw relationships among objects in various situations. Associations can be made by various attributes (size, shape, color, use, etc.), by order in a sequence, by position in a pattern, etc. In math there are certain techniques used to represent these relationships. This exercise has the students use their skills of sorting by attribute together with Venn diagrams to represent their groupings.

Have the students cut out pictures from various magazines (it may save time and classroom frenzy to have students do this for homework). They should organize their pictures into three identifiable groups with about four to six items in each group. They should then make a collage by scrambling the pictures together and gluing them to the paper. Now have them draw out a Venn diagram representing their collage (it may help to model this procedure at the board with a few classroom items). They should not share their work at this time although they will naturally want to exhibit the pictures they have found.

Next have them trade papers with someone across the room. Their job is to examine their classmate's collage and recognize some groupings. They must then draw a Venn diagram to represent their reasoning. After all the collages have had two Venn's done, have a discussion comparing the results. By posting them on the board you can display them and generate non-threatening discussions. Point out that if someone can recognize a logical relationship among the pictures in the collages it should not be considered wrong even if it deviates from the author's Venn. The goal is not to exactly mirror the original Venn but to polish students' ability to reason logically.

26	Uses Venn Diagrams to represent complex relationships (sets, subsets, etc.)

Jokers Wild

Grade Level: Middle

MATERIALS: Deck of cards (including jokers)

ORGANIZATION: Whole class activity

PROCEDURE: Explain that in logic there are certain techniques used to illustrate relationships among various objects.

One of these techniques is the use of Venn Diagrams to show objects that share or do not share attributes. Venn groupings can show the following characteristics:

Universal set:	All of the object with which you are working
Union of sets:	Combining all of the members of two or more sets into one set
Intersection of sets:	All the common objects shared by two or more sets
Subset of a set:	A set formed from items within another set
Disjoint sets:	Two sets which have no common objects (no intersection)

Tell the students that they must do this activity as quietly as possible because they will be listening to instructions as to where to move in the classroom. Give a playing card to each student and begin by asking all members of the universal set to stand up (everyone). Have the subset of HEARTS move to one corner of the room (a different corner for each suit). At this point remind student that in terms of card suits, they are in four disjoint sets; however, since each suit has the same card values, there are many set intersections which can be made.

Ask a student to suggest an intersection (e.g., all facecards, all cards less than 5, all cards with a letter, etc.) and have these subsets move to the middle of the room. Depending upon the sophistication of your class there a many possible sets and unions to be formed. Once they have finished this activity, have them return to their desks and draw the various Venn diagrams which they formed. Have students share their diagrams.

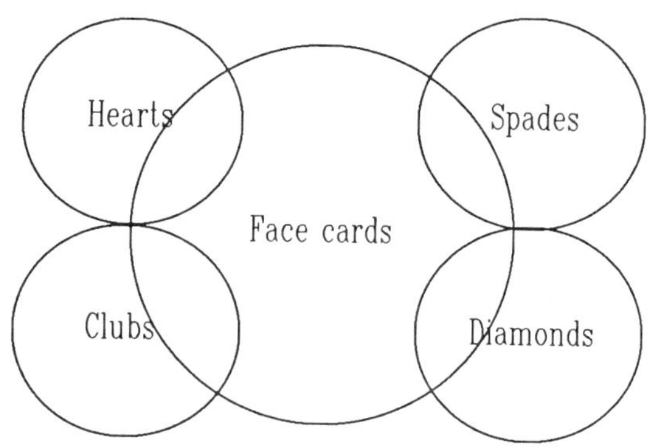

26	Uses Venn Diagrams to represent complex relationships (sets, subsets, etc.)

Letters are Venns?

Grade Level: Upper

MATERIALS: Index cards (or smaller cards of some type) numbered from 1 to 15, markers

ORGANIZATION: Individually or in teams of two, three, or four students

PROCEDURE: This is a difficult concept for students to grasp but if they can understand this activity, they have an excellent understanding of Venn Diagrams.

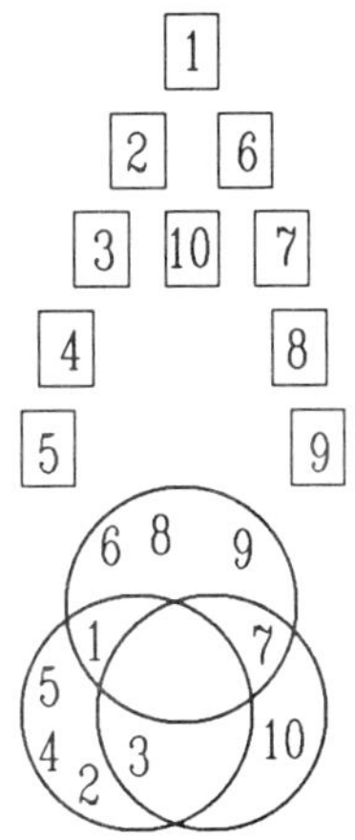

Students will be using the letters of the alphabet to find intersections of sets and will then create a Venn Diagram for each letter of the alphabet. The solutions are shown on the next page. For explanation, we will use the letter A.

Begin by arranging a set of cards (as shown) on the chalkboard or overhead projector. Have students imagine that each "leg" of the letter A is a different set. Have students record the sets: left diagonal {1,2,3,4,5}; right diagonal {1,6,7,8,9}; and cross piece {3,10,7}. Now ask students to join these sets in a Venn diagram so that card 1 is shared by the left and right diagonal; card 3 is shared by the crosspiece and left diagonal; and card 7 is shared by the cross piece and right diagonal. The answer is given.

Divide the class into groups so that students who understand the concept are grouped with students who are having difficulty. At this point there are two ways to proceed. You might continue with the procedure by displaying each letter configuration on the overhead and keeping the class working within their groups but at the same pace, or you may wish to hand out a xeroxed copy of the reverse side of this page (you will need to omit the solutions) and have each group work at its own pace.

Students will begin to find patterns in the letters and will find these patterns create similar Venn Diagrams.

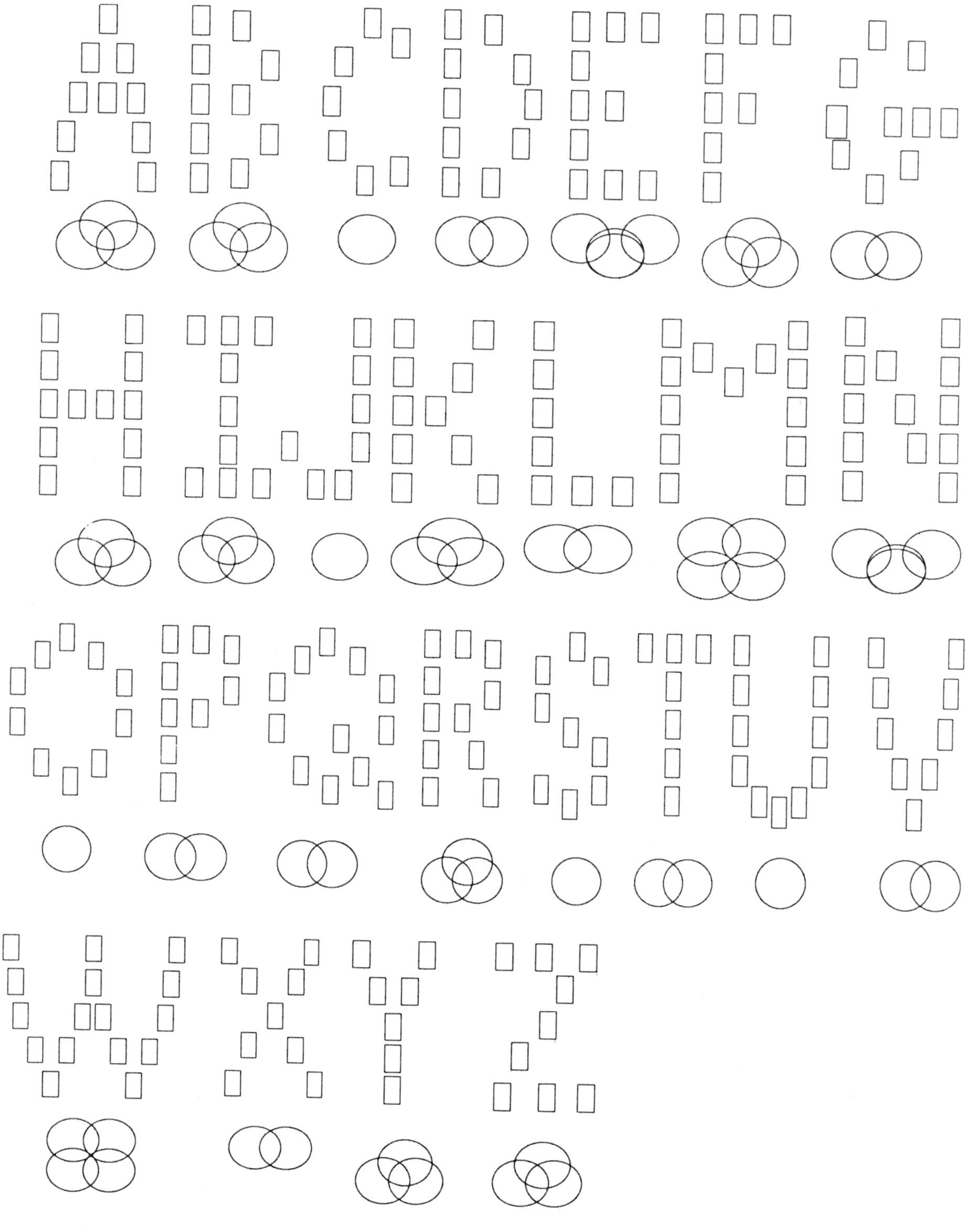

27	Identifies relationships among three or more objects according to 3 or more attributes

Scavenger Hunt

Grade Level: Middle/Upper

MATERIALS: Scavenger hunt cards (as shown below), sample items to collect

ORGANIZATION: Cooperative teams of three to six students

PROCEDURE: Tell the students that they are going to be walking around the school or school yard and collecting objects that in some way can be compared to a given item (as an example, we will use a tennis shoe). The teacher should model this activity in the classroom by searching the room and finding three items that have shared attributes with a shoe -- these might include a black pen (color), a stapler (basic shape), and a jacket (cloth).

Discuss how these attributes relate to the shoe (color, shape, and composition) and ask students if they can find other items in the classroom that share a different attribute with the shoe. List these items and relationships as students identify them, but accept only ideas that create a new relationship (i.e. size, function, texture, etc.) Once a list of five to ten items has been generated tell students that these are typical attribute categories. Discuss whether or not any particular item would have these attributes (yes).

Tell students that they are going to go on a scavenger hunt but rather than finding a list of "things," they are going to to find items that share an attribute with one item. A sample scavenger hunt list is provided. Once students have collected their samples, discuss the findings with the entire class.

Item: ____________
Attribute
Color
Size
Shape
Texture
Composition
Function
Other

27	Identifies relationships among three or more objects according to 3 or more attributes

A Missing Link (Part 2)

Grade Level: Upper

MATERIALS: Various items for attribute identification -- e.g. apples, candles, scissors, etc.

ORGANIZATION: Teams of two, three, or four

PROCEDURE: Begin by having the class identify three or more attributes of a selected item (e.g. an apple). Prompt this response by listing several possible attribute categories on the board (size, color, shape, uses, component parts, or texture, to name a few).

Using the apple as an example, students might generate attributes (see below) such as "red," "spherical," and "smooth." Ask students if they can think of other objects which match these attributes.

As you discuss this, allow students to come to the realization that the more attributes which are identified regarding an object, the fewer objects will match the attributes.

List several more attributes of the apple to form an "attribute chain" on the chalkboard or overhead.

red -- spherical -- smooth -- shiny -- edible -- size of a child's fist

Given this chain, ask students to change one attribute and then to try to identify an item which matches this new "chain." At this point you might divide students into groups to work on generating a group chain. Groups will then make replacements. You might also ask students to do this individually or with the help of parents as a homework assignment.

ATTRIBUTE CHAIN	IDENTIFIED ITEM
RED/SPHERICAL/SMOOTH/SHINY/EDIBLE/SIZE OF A CHILD'S FIST	APPLE
RED/SPHERICAL/SMOOTH/SHINY/EDIBLE/SIZE OF A MARBLE	HARD CANDY
RED/SPHERICAL/SMOOTH/SHINY/NON-EDIBLE/SIZE OF A MARBLE	RUBBER BALL

27	Identifies relationships among three or more objects according to 3 or more attributes

Not Just a Bug!

Grade Level: Upper

MATERIALS: A collection or pictures of insects

ORGANIZATION: Groups of two, three, or four students

PROCEDURE: In this lesson, students will be analyzing and categorizing insects by various attributes.

Begin by dividing the class into cooperative groups of two, three, or four students and give each group three insects (pictures or real). Have students study the specimens and see if they can find similarities and differences. They should search for attributes such as -- SIZE, COLOR, SPOTTED, STRIPED, WINGED, LENGTH OF LEGS, ANTENNAE, SHAPE AND SIZE OF HEAD, ABDOMEN, THORAX, etc. Ultimately, they should identify the attributes of each insect which set it apart from the others. Have them list the differing attributes on index cards. Have them arrange the insects and index cards as if it were a display.

Allow students time to circulate the classroom to view the groupings of classmates. You might also wish to ask each group to give a brief talk describing why they grouped their insects as they did.

As an extension, have each student or group of students write a paper elaborating on their ideas of the relationships. The papers can be read aloud with a discussion period after each.

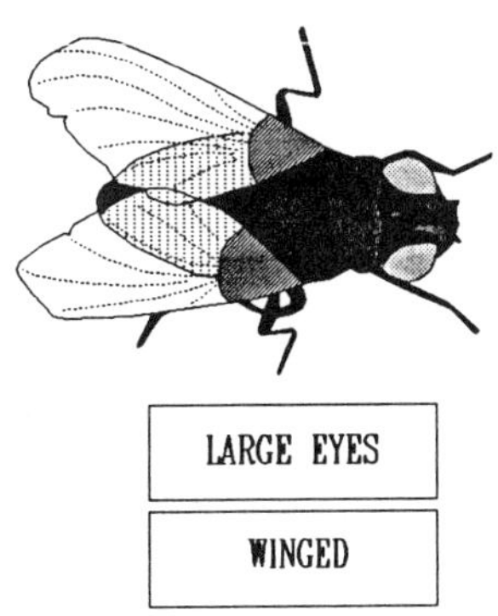

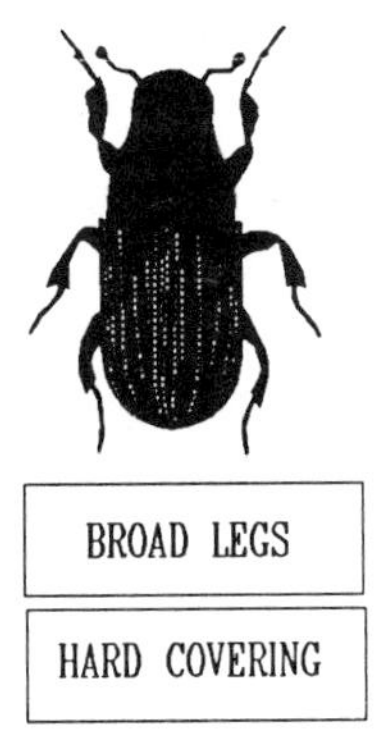

27	Identifies relationships among three or more objects according to 3 or more attributes

Happy Holiday
Grade Level: Upper

MATERIALS: No special materials are necessary

ORGANIZATION: A whole class or small group activity

PROCEDURE: This lesson takes the concept of related attributes to an abstract level of students finding relationships between the purpose, background, and traditions of various holidays

Begin the lesson by generating a list of holidays with the students. Include anything from religious holidays to national holidays to personal holidays such as birthdays and anniversaries. A list might include: Christmas, Easter, Hanukkah, 4th of July, Veteran's Day, Halloween, New Year's Day, Memorial Day, Birthday, Valentine's Day, President's Day, Mother's Day, Father's Day, Labor Day, Anniversaries, and Martin Luther King Day.

Discuss with the students as to how these days are similar and how they differ. Their task is to take the days on the list and create categories. Some possible categories include: days honoring people, days on which gifts are given, days the world celebrates, days celebrating a religious event, days of personal importance, and days celebrated by the country. There are many more categories as well.

Once students have categorized these days to their own satisfaction have them create a poster or chart which shows their organization. This is a very open-ended project and students may create numerous ways to present this information. The display is secondary to the logic involved in analyzing the purpose of each day and classifying the inter-relationships of each day.

28 Identifies complex relationships among objects according to position in a pattern or sequence

Mega-toe

Grade Level: Upper

MATERIALS: Paper and pencil

ORGANIZATION: Groups of two

PROCEDURE: In this activity, students will be looking at game situations in order to discover something about the nature of certain patterns.

Some students may already be aware that the game of tic-tac-toe will always result in a draw if the players plan their moves in advance. With this in mind, you will probably need to begin by saying that there is a secret about the game and that if anyone knows it, they should help keep it a secret until the whole class has had a chance to discover this fact for themselves.

Allow students to play tic-tac-toe for five to ten minutes; then discuss the results of the contest. Ask various teams how many games each person won; if it mattered who started; if it mattered where the first mark was made; etc. Eventually let someone disclose the "secret" that the game is a "no win" game. Allow someone to challenge an "expert" if there are any doubters.

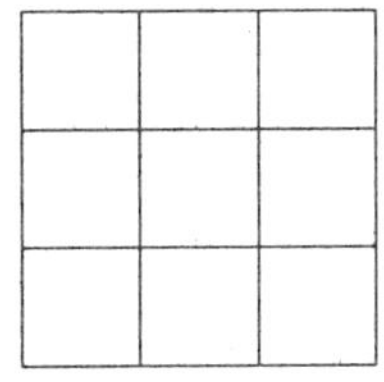

The next step is to announce that they are going to play another game called "mega-toe." Draw a grid on the board which is five squares by five squares . Tell students that in order to win they must complete a straight line either diagonally, vertically, or horizonally. Let them play the game for ten to fifteen minutes and have them write out their results, stress that they must include reasons why they got the results they did. Next, have a discussion about the results, eventually coming to the conclusion that it is another "no win" game. Have students explain the reason for this "no-win" situation.

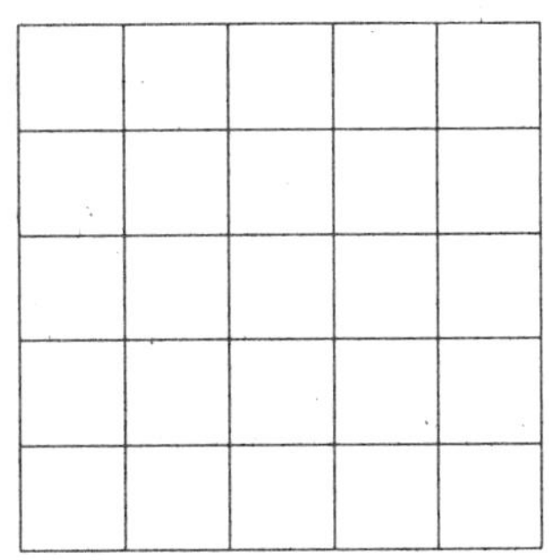

The 5 x 5 grid is actually a tic-tac-toe grid expanded by one row on all sides. The inherent traits of the game have not been changed, only its appearance . Although it has no routine pattern, it does have a " random" pattern in that the result is always a draw.

As an extension, have students create their own "no win" games; change certain "no win" games so that it is possible to win; or change certain common games so that they become "no win" games.

28	Identifies complex relationships among objects according to position in a pattern or sequence

Any Ding-a-ling Will Do

Grade Level: Upper

MATERIALS: 8 canning jars and 8 more assorted jars, water, a striking instrument, masking tape

ORGANIZATION: Small group or whole class activity depending upon the number of jars

PROCEDURE: This lesson is based on the experiment using various amounts of water in a jar or glass to see how it affects the pitch when it is struck. Have students create logical conclusions regarding the size of containers and the amount of water in the container.

Begin by giving each group (or one set for the entire class) a set of 8 jars. All eight should be the same size and shape. Pour water in jars and arrange them so that they form a musical scale. Have students create a syllogism which explains the relation of the amount of water to the pitch of the jar (e.g. a full jar of water makes a low sound; and an empty jar of water makes a high sound; therefore, a half-filled jar will make a sound between a high and low sound).

In the second section of the lesson, students will introduce a new factor -- the size of the jar. Give each group eight jars of different shapes and sizes (but no water and nothing with which to strike them) and have them arrange the jars in the order that they think they will make low to high sounds. If students think that they will have to add water to certain jars, have them mark this imaginary line with masking tape.

Before testing their theories, have students write statements which explain their logic, i.e. smaller jars will have a higher sound than large jars; thin glass will make a higher sound than thick glass; a wide opening will make a lower sound than a small opening, etc.

Give students water and strikers to see if their statements are true. Have them readjust their statements and see if they can write syllogisms regarding jar size, glass thickness, or amount of water.

28	Identifies complex relationships among objects according to position in a pattern or sequence

Gearing Up

Grade Level: Upper

MATERIALS: Copies of next page, gear assortments from various toys sets (if available), scissors, a ten speed bike

ORGANIZATION: A whole class activity

PROCEDURE: This is a thought provoking lesson involving both the logic of patterns and the theory of gears. It is best if a ten speed bike can be used in the classroom to demonstrate the value of gear patterns.

Students will need to understand some vocabulary to enjoy this lesson: drive gear - the gear which is turned by the power source (pedals), slave gear - the gear which is turned by the drive gear (wheel), gear teeth.

Have students cut out the gear models on next page. Have them complete the chart (shown below) by experimenting with the three gear sizes -- writing down the number (or fractional part) of revolutions in each gear combination (the chart is complete for the teacher's use).

Once this information is gathered, divide the class into groups of four and have them discuss and then write statements which they think will be true -- i.e. If the drive gear is small and the slave gear is large the bike will be easier to peddle because...; if the drive gear is large and the slave gear is small, the bike will be hard to peddle because... Once students have shared their statements, allow students to experiment with the various gear arrangements on the ten speed bike to either prove or disprove their deductions.

	# of teeth	Rev. around large gear
Gear 1 (large)	52	-
Gear 2 (medium)	40	1 1/4
Gear 3 (small)	28	1 7/8

As an extension, you might have student bring various geared apparatus from their homes, such as slot cars, pencil sharpener, can openers or egg beater. Have student discuss how the gears make these mechanisms work for effectively.

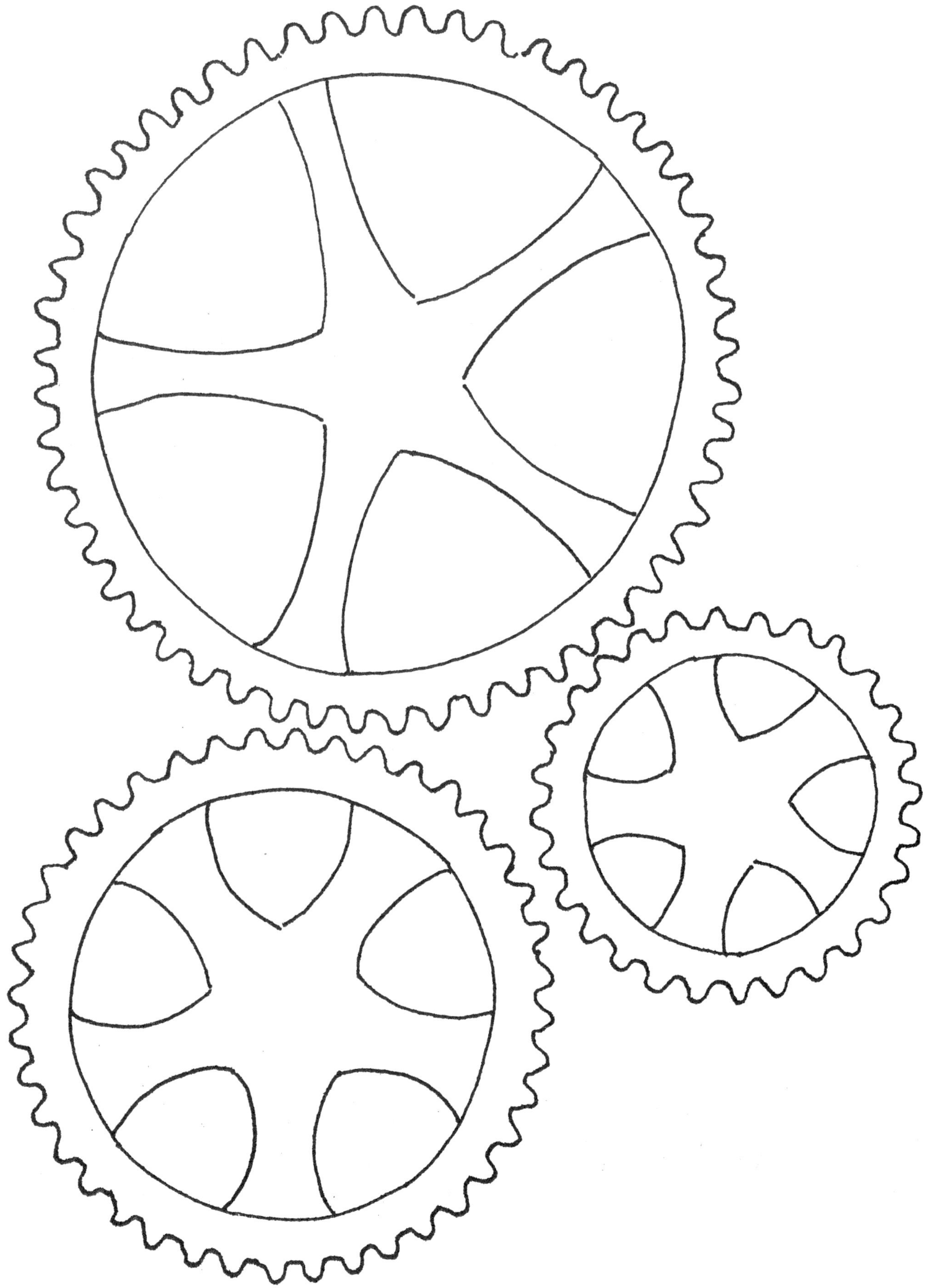

28	Identifies complex relationships among objects according to position in a pattern or sequence

Can You Draw What You Don't See?

Grade Level: Upper

MATERIALS: Sample cube as shown on next page, scissors, scotch tape

ORGANIZATION: Individually or in teams of two students

PROCEDURE: In this lesson, students will be using their spatial relation abilty to draw various views of cubes.

Begin the lesson by giving each student one of the cubes as shown ion the next page. Have students cut out the cube and tape it together.

Have the student place the cube on the table so that one face can be seen. Using one of the "model cubes" (unfolded) see if students can draw the design on the opposite side of the cube -- it should be drawn so it is facing the right direction. Repeat this task several times until students can complete the task successfully.

Some students may be able to complete only this much of the activity; however, for students who are able to do the drawing, have them also try to draw the bottom side of the cube.

The next step is to have students use two cubes side by side and try to draw the back side and bottom side of the two cubes. For students who can handle this, see if they can draw the two sides which touch one another. The teacher should note that there is a logical pattern in the cubes, the four squares on the vertical of the "model" cube are diagonal patterns, the other two squares are vertical/horizontal patterns.

As an extension, you might want to have students create their own cube patterns or do the same activity with triangular prism shapes.

29	Develops relationships among objects and draws conclusions based upon certain qualities or conditions

Try This Angle!

Grade Level: Upper

MATERIALS: Construction paper, protractors, pencil, enlarged copies of triangles on this page

ORGANIZATION: Individually or groups of two, three, or four

PROCEDURE: In this lesson, students will identify patterns found in triangle shapes. These should include: three corners, three sides, three angles, and most importantly, 180 degrees.

Begin by handing a copy of the triangles to each group and tell them to write four attributes shared by all five triangles (isosceles, equilateral, scalene, right, obtuse). The first three attributes will be easy for students to identify but the fourth attribute will require the use of a protractor and totaling the three angles. It will take time and some leading questions provided by the teacher.

Once students have discovered the fourth attribute, pose two more problems for their solution. First, have them discover shared attributes of various four or five-sided figures. Second, ask them to try to construct a magical triangle with angle C twice the size of angle B, and angle C three times the size of angle A (30-60-90 degree triangle).

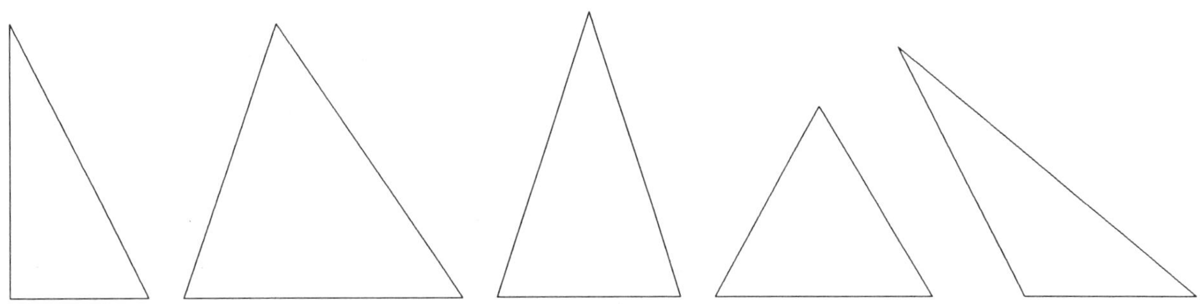

right triangle　scalene triangle　isoceles triangle　equilateral triangle　obtuse triangle

29	Develops relationships among objects and draws conclusions based upon certain qualities or conditions

Who's Coming to Dinner

Grade Level: Upper

MATERIALS: Clue cards, table with chairs, pencil

ORGANIZATION: A whole class activity eventually dividing into groups of two

PROCEDURE: Begin by duplicating the clue cards and set up a table with eight seats according to the diagram below. Select eight students to play the characters on the clue cards. Tell them they have been invited to a masquerade party and must find their seats according to the clues on their cards.

Have each student take a clue card to determine their character. Students should work together to solve the problem. After the students have solved the problem have them read the clues again to see if all facts are true. Have each group of two then plan a Halloween Party with clues for a seating arrangement of eight people. Each group will select eight people to go to the front of the room and seat themselves according to the clues.

Read this clue aloud. You may not pass it to anyone else.
The King sits at the head of the table.

Read this clue aloud. You may not pass it to anyone else.
The Prime Minister sits to the right of the king.

Read this clue aloud. You may not pass it to anyone else.
The prince sits opposite the princess.

Read this clue aloud. You may not pass it to anyone else.
The Ambassador is to the right of the prince.

Read this clue aloud. You may not pass it to anyone else.
The princess sits between the prime minister and the senator.

Read this clue aloud. You may not pass it to anyone else.
The queen sits to the left of the president.

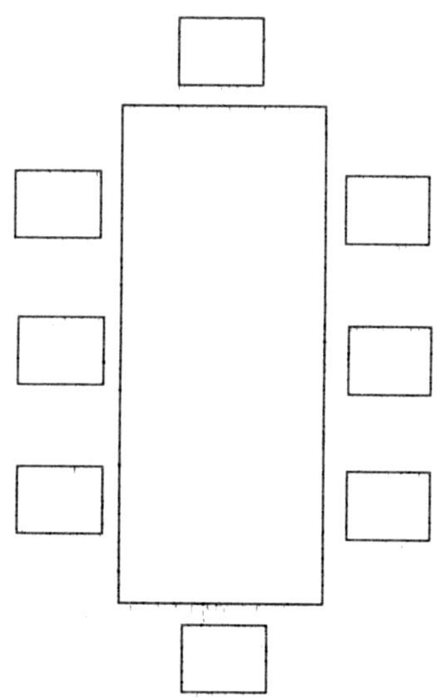

29	Develops relationships among objects and draws conclusions based upon certain qualities or conditions

Creative Machines

Grade Level: Upper

MATERIALS: Paper cups, flex straws, paper clips, rubber bands, tape, magnets, and string 18 inches long. Paper and pencil to make sketches

ORGANIZATION: Groups of three or four

PROCEDURE: Begin by discussing with the class the idea that machines are made up of various parts which have been designed to work together to accomplish a specific task. The way the parts are put together involves logically identifying the various capabilities of each component part and how it interrelates with the other parts.

Explain to the class that their task will be to use the materials provided to construct a machine capable of performing a simple task (e.g. to scoop or lift various materials). Each group will be responsible for making detailed drawing of their machine which adequately illustrates how the different parts function together to perform the task.

Have the class sketch their ideas before actually constructing their machines. After they have submitted an initial prototype drawing for "approval" by the engineering department (the teacher), have them construct a working model. Allow them the opportunity to correct mistakes or refine their machines. By using cooperative groups, students that are predominantly manipulative learners will learn to share their ideas with others.

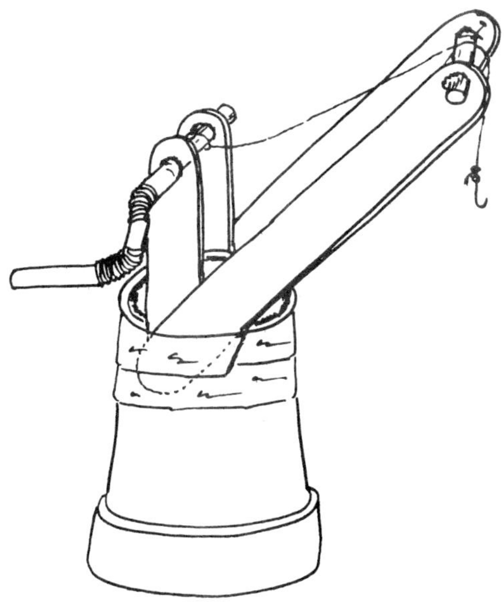

We suggest that you have students create the ideas for their machines, but if students become stumped, you might refer them to a unit on machines in the science book or encyclopedias. Some other simple machines might include a crank crane, a catapult, or a scoop of some type.

29	Develops relationships among objects and draws conclusions based upon certain qualities or conditions

Fried and True Research

Grade Level: Upper

MATERIALS: Paper and pencil for taking notes

ORGANIZATION: Individually or groups of two

PROCEDURE: Begin by explaining that in order to make correct decisions, people need to be able to make observations and recognize relationships from their results. This is one of the things that has made McDonald's restaurants so successful; the ability to study an area and reach a valid conclusion based on their observations. They know that they will sell the most food in an area where there is heavy foot or car traffic.

Have your class imagine that the student council has hired them to do a demographic study of the school. They have purchased two newly developed french fry machines. They have decided to place one in the school cafeteria but need to know where to install the other machine for the greatest profit. The student's job will be to observe the traffic patterns of the student body to find out where the traffic is the heaviest.

Have the students begin by drawing a map of the school. Next have them discuss or write out where they think the best place or places would be to install the machine. The key element in this process is the reasoning. Now let them do the research to see if their supposition is correct. You may decide to have all the students gather their own data or divide the areas and have students cover designated areas. In either event, you should have the students research the traffic patterns at different times.

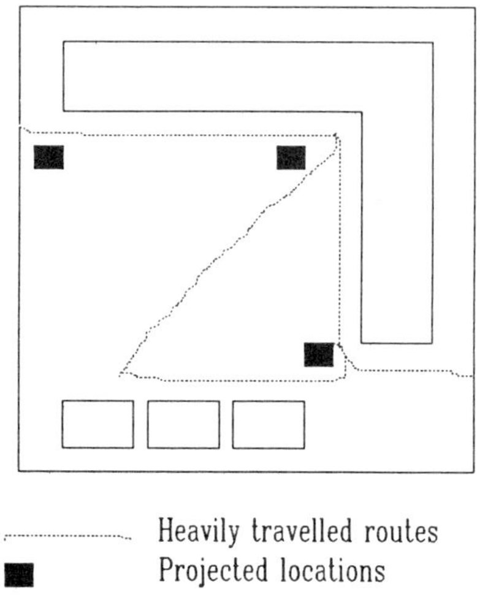

As an extension, have them do research to decide where extra trash cans should be placed (i.e. at what distance from the machine and along which routes).

30	Solves complex problems based on disqualification of available options

Mystery Shoeboxes

Grade Level: Upper

MATERIALS: Mystery Shoeboxes for each team

ORGANIZATION: Teams of two to four

PROCEDURE: You must first construct several "mystery boxes." Do this by taking an object (tennis ball, golf ball, pebble, small fork, scissors, etc.) placing it inside the box, coding it for identification later, then sealing the box with tape. The students will be trying to surmise what is in the box without looking. It may be best to start with objects that share a certain trait, purpose, or location. Inform the students that the objects are related (e.g. household items, tools, sports, etc.). Later, as students become more practiced, let them experiment for awhile on their own and then compare notes to try to find the various objects' interrelationships.

Explain that in logic, people frequently gain much information about what an object is by finding out what it isn't. Further, explain that by recognizing certain attributes or qualities of an unknown object, the identity of that object eventually becomes more clear. Take suggestions for some experiments they might make: shaking, tilting, dropping, etc. By listening and feeling the weight shift they will be able to start eliminating possibilties. Discuss a method of record keeping so that they will consistently categorize their information.

Once they are comfortable with the procedure, distribute the boxes and have them start to experiment. After they have had a chance to explore the possibilties have them write down what they think the objects in the box might be. Have them pass the boxes on to the next team.

After each team has had a chance to experiment with each box and has written down a conclusion, have a class discussion, comparing notes, procedures, results, and conclusions. Stress that the goal is not necessarily to identify exactly what is in each box, but rather to examine how logical thinking works and to reach a "valid" conclusion that can be supported by data. Any answer that fills these criteria should be considered correct.

Some possible permutations are to give the class some clues (the objects are all blue, one is blue, one is red...; they all cost less than a dollar, etc.) or let them see an array of items (perhaps 12), and place nine in their respective box (this will give them a frame of reference without giving away the solution. Another alternative is to have the students eventually draw pictures of what they think the object's shape is. Then, post some photos on the board and have them identify which one goes with which box (you might have extra pictures to make it more challenging).

30	Solves complex problems based on disqualification of available options

Human Matrix Grids

Grade Level: Upper

MATERIALS: Name tags with removable cards, markers, printed matrix logic puzzles

ORGANIZATION: Small groups or a whole class activity

PROCEDURE: There are many logic matrix materials available and in this lesson we propose a format to use these materials in a new and unique way.

Begin this activity by giving a group of students name tags and markers. Using the sample problem, have students transfer this information to a name tag and wear it throughout the school day. It will be each student's job to attempt to talk with these students and figure out the solution to the problem.

Food, Food, Food!

Juanita, Eduardo, Sue, Sharon and Andy each have a special favorite food. No two of them have the same favorite. Which child goes with which food?

1. I'm Juanita and my favorite is not pizza.
2. I'm Eduardo and I hate pizza and hamburgers.
3. Someone really likes hot dogs.
4. I'm Sue and I like french fries.
5. I'm Sharon and I'm allergic to onion rings.
6. I'm Andy and I like the food to which Sharon is allergic.

Given this problem (and other matrix problems) some students will feel successful in just identifying how many students and how many foods are involved. Other students will want to complete the entire solution.

As an extension, you may want to have students omit the name tags and assume the role of each character, parroting statements for classmates as questions are proposed.

30	Solves complex problems based on disqualification of available options

Treasure Hunt

Grade Level: Upper

MATERIALS: 1" grid paper for each student

ORGANIZATION: Teams of two

PROCEDURE: Explain to your students that the goal of logic is to enable people to come to conclusions which are based upon information. This makes the conclusions reached by people more sensible and more accurate. This exercise involves correlating information in order to locate a hidden treasure/object. This is somewhat similar to a game some students may have played called Battleship.

Begin by drawing a ten by ten grid on the chalkboard, similar to that shown. Explain that they are to discover the hidden location of the lost treasure by gathering information about compass points. For example, if the treasure is hidden on square A-1 (uppermost left corner) it would be located in the northwest. If a team starts their guess by saying F-6 (center of the grid), you would give them the clue that the treasure is west (or north) of their camp. You should only give ONE direction for each question (i.e. either north or west). The choice of which clue to give can be discussed with the students. After they have gone through the game once, have them discuss the types of strategy that will help them find the treasure most quickly.

After they have become familiar with the process, have the teams use their grids to challenge each other. Each team hides a treasure on their "map" (they can write it on a separate piece of paper or use the grid and keep it hidden from the other team). They then quiz the other team about locations, taking turns, of course. For closure, have them write out or discuss their results, being sure to cover the reasoning processes they went through.

	1	2	3	4	5	6	7	8	9	10
A										
B										
C										
D										
E										
F										
G										
H										
I										
J										

30	Solves complex problems based on disqualification of available options

Playing With a Full Deck

Grade Level: Middle/Upper

MATERIALS: Decks of cards, grid check-off sheets (next page)

ORGANIZATION: A whole class activity, dividing into groups of four

PROCEDURE: In this activity, students will use a deck of cards to present logic problems to classmates.

Begin by demonstrating the game to students. Randomly select two cards from the deck. Ask students to pose questions which will help them to guess the cards (number and suit). Sample questions might include: "Are the cards red?" "Are the numbers greater than 5?" "Are they face cards?"

To each question, you will respond with a "0" if neither card fits the description; "1" if one of the cards fits a description; and "2" if both cards fit the description. As students ask questions, ask them to give the logic as to why the question is a "good" question. With each question's response they will check off the "learned" information on the accompanying grid. There are two rules for questions. They cannot use the conjunctions and, but, or and they cannot guess an actual card name until all but eight of the grid squares on their grids are crossed off.

There are many approaches the teacher can take in this demonstration game. If your class is new to logic puzzles, you might explain step by step what information they have gathered by each response. A more sophisticated class might be left to their own devices to try to solve the puzzle. Students might be grouped in cooperative teams to discuss approaches to solving the problem.

A sample game might include the following questions and responses:

Students	Teacher
	Selects two cards (3 of hearts and 7 of clubs)
Is it red?	"1" (one red and one black card)
Does it have a letter?	"0" (both cards are between 2 and 10)
Is it even?	"0" (both cards are 3,5,7, or 9)
Is it a heart?	"1" (one card is a heart)
Is it a spade?	"1" (one card is a club)
Is it greater than 6?	"1" (one card is a 3 or 5 and one is a 7 or 9)

Once students have fewer than 8 grid squares which are not crossed off, they can then make a logical guess as to the cards held by the teacher.

	♣	♥	♠	♦
Ace				
2				
3				
4				
5				
6				
7				
8				
9				
10				
Jack				
Queen				
King				

30	Solves complex problems based on disqualification of available options

Mystery Bags

Grade Level: Upper

MATERIALS: A bag containing a mystery object (one for each group)

ORGANIZATION: Groups of two, three, or four

PROCEDURE: This is a logic guessing game in which groups of students try to guess one another's "Mystery Item."

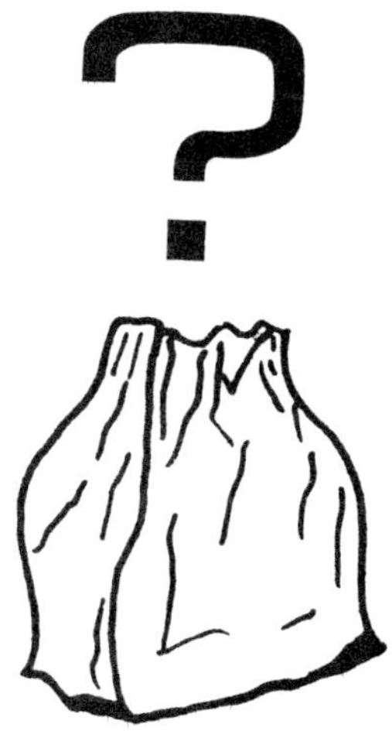

In the beginning part of this activity students will be required to ask only yes and no questions about an unknown object (the object in the bag). From their information they must figure out what is in the bag. Have the groups keep notes as they ask questions; they should keep track of all the answers since this will increase the information they have -- it will also encourage them to pay attention. Once a group thinks they know definitely what the object is, allow them to guess when it is their turn. Allowing them to bet fictitious points (i.e. giving them a bankroll of 50 points) usually heightens their interest and prevents frivolous guessing.

After the object has been guessed, have a discussion about the different strategies groups used to ask questions and organize their information. They should be encouraged to use strategies such as focusing on one attribute at a time (i.e. size, color, weight, possible uses, shape, etc.) or formulating a hypothesis/guess based on acquired information.

Once the students have acquired some skill in asking appropriate questions and organizing their information, give each group a bag with an unknown object. Each group is to examine the object in their possession so that they will be able to answer questions, but they should not let the other groups see what their object is. Allow each group an opportunity to go to the front of the class to answer questions from their classmates about the bagged item.

They should record their answers and start to modify questioning strategies. Then moving in a clockwise direction, designate the next group to be questioned by the class. Continue to designate groups to be questioned until each group has been "interrogated" by all the other groups. This finishes one cycle. Keep repeating this cycle until one of the groups is positive that it can correctly identify the contents of all the bags.

It is possible that none of the groups will be able to do this in one sitting. In order to extend the inquiry it may be necessary to have some reward not only for the first group to get all the objects correct but also to the group which has an object which none of the others can identify. This should prevent friends from divulging information.

31	Demonstrates an understanding in using the words: IF...THEN, IF...NOT, and IF AND ONLY IF

Making the Circuit

Grade Level: Upper

MATERIALS: Highway road maps duplicated for each student (or group)

ORGANIZATION: Individually or in teams of two or three

PROCEDURE: Road maps provide an excellent means for teaching children about "IF..THEN, IF...NOT, and IF...AND ONLY IF" statements. This lesson will be based upon the map displayed below. We suggest you do this lesson in two phases. First, use the same map for the entire class to explain the concept of the lesson. Second, let students select their own maps to add a personal touch to the lesson concepts.

Introduce the lesson by handing out a map (use ours or your own) and asking students to describe how they would get from Pueblo to Colorado Springs. Most will identify highway 25 (85/87). Have students say this as a statement: "IF you want to go from Pueblo to Colorado Springs, THEN you can take highway 25."

Ask students to describe how they would go from Pueblo to Buena Vista. This is less obvious and several responses are possible. Once again have students state this as an "IF...THEN" statement.

Given the map, have students see if they can create an IF...NOT statement (i.e. IF you want to go from Pueblo to Colorado Springs on highway 5, you can NOT go to Canon City). Students should create several of these statements.

Finally have students create "IF AND ONLY IF" statements. "You can go from Pueblo to Colorado Springs on highway 25 IF AND ONLY IF you go through Widefield" or "You can go over Currant Creek Pass" IF AND ONLY IF_you go on Highway 9.

Based on these types of statements, have students write their own "IF" statements. The product of this activity is not the important factor, the logical thought of creating statements using specific language is the purpose of the lesson. Once students have completed this portion of the lesson, let them work on their own map area to present to the class as a poster or display.

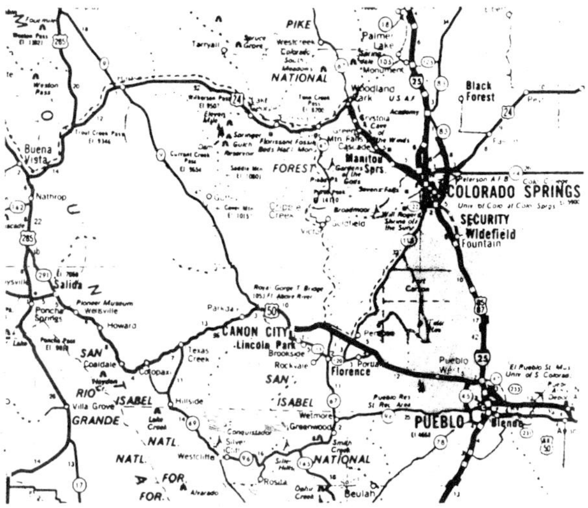

31	Demonstrates an understanding in using the words: IF...THEN, IF...NOT, and IF AND ONLY IF

Logic Rebuses

Grade Level: Upper

MATERIALS: Scissors, markers, art paper, glue, magazines

ORGANIZATION: A whole class activity or teams of two

PROCEDURE: In this lesson, students will be creating relationships using the "IF..THEN" statement format. In this activity, each student is to create a rebus (pictorial sample as shown below) to demonstrate the statement.

Begin by writing samples of IF...THEN statements on the board and explaining the relationships involved. Examples of these statements are: IF yellow vegetables are rich in vitamin A , and crookneck squash is yellow, THEN crookneck squash is rich in vitamin A; or, IF only birds can fly, THEN crows are birds.

Give students an opportunity to generate their own statements (statements should be kept to the students rather than shared). Have students search through magazines to find pictures to make rebuses out of their logic statements. Students can share their rebuses to see if classmates can solve the logic statement.

As an extension, you might have students experiment with the use of "IF...NOT" and "IF AND ONLY IF" statements in the student's rebuses.

31	Demonstrates an understanding in using the words: IF...THEN, IF...NOT, and IF AND ONLY IF

Assembly Line Sorting

Grade Level: Upper

MATERIALS: Cards for attribute words, string, markers, a bag with many items for sorting or pictures cut out of magazines

ORGANIZATION: Begin with one demonstration group then a whole class activity

PROCEDURE: In preparation for the lesson prepare sets of cards with the following attribute categories: shape, size, color, function, composition, and cost. You will need to prepare enough sets so that each student in your class has a card. A nice touch is to put a string on the card so the student can wear it around his/her neck.

Write one of the following adjectives on the back of each shape card: round, flat, rectangular, uneven, and cylindrical. On the back of the other cards write:

size*	very small, small, medium sized, large, very large
color	pastel, dark, bright, multi-colored, one-color
function	used by men, used by women, used by children, used by students, used by animals
composition	plastic, metal, wood, manmade, natural materials
cost	less than $2.00, less than $5.00, less than $10.00, more than $2.00, more than $5.00

Front Side:	SIZE	SIZE	SIZE	SIZE	SIZE
Reverse Side:	Very Small	Small	Medium	Large	Very Large

Select six students to be the demonstration group and have them come to the front of the class. Give each of them a different attribute card, allowing everyone to see the attribute but no one but the individual to see the adjective.

*You may want to make this "relative" to a softball or yo yo.

Tell the class that this is the inspection team for a new set of products known as "burbles" (create some type of nonsense word to draw the studentS' attention). Hand an object to the first person in line and have him analyze it. If it matches his adjective he announces that it has the "shape" of a burble and hands it to the next student to check his attribute. If it does not match a student's adjective, the item is rejected and placed back on the table.

Only those items that pass the scrutiny of each team member make it through the line to be labeled a "burble." Once an item has made it through the line, have the remainder of the class guess the adjective under each attribute sign. If students cannot guess the correct attributes, send other items through the line for scrutiny.

Once students understand, have the students form lines radiating from a central table in the room (like spokes of a wheel) and place numerous items for consideration on the central table.

Give each student an attribute card and give the students time to sort through the items on the table. The last person in the line should keep all of the items which have met that group's criterion. Once finished, have each group share their "product" and let classmates guess the attributes which each person use for sorting.

31	Demonstrates an understanding in using the words: IF...THEN, IF...NOT, and IF AND ONLY IF

Let There Be Light

Grade Level: Middle/Upper

MATERIALS: Light bulb on extension cord, various materials (paper bag, large can, glass jar, translucent container, etc.)

ORGANIZATION: A whole class activity

PROCEDURE: Begin by explaining that in the study of logic, using a certain vocabulary helps keep information in order. It also emphasizes the differences and similarities of given information. Review definitions of the words transparent, translucent, and opaque.

Introduce the idea that some material will allow light to pass through. Display the various items listed above and ask the students which will allow light to pass through. Have them make statements (either written or verbally) such as: "IF AND ONLY IF a material is transparent, will light shine through," or "IF a material is opaque, THEN light will not shine through." Test each statement and have the students revise those which are incorrect. Discuss why some materials transmit light and some do not.

Have them extend these class-observation statements to discuss objects outside of the classroom. They should sort these statements into appropriate categories and explain their reasoning using the vocabulary developed in this lesson. You may have students bring objects at a later time to test their reasoning.

An extension to this activity is a science discussion of absorptions and diffusion of light.

Glossary

Deductive Logic	Reasoning from a known principle to an unknown or from a generalization to the specific concept.
Extrapolate	A projection of future expectations based upon past results. It differs from interpolation in that extrapolation is a prediction or extension of given information.
Inductive Logic	Reasoning from specific facts to a generalized statement.
Interpolate	To insert in or among given information. It differentiates from extrapolation in that interpolation is the completion of given information within a specific set.
Pattern	A set of items which repeat in the same order.
Sequence	A set in which there are various components which do not repeat as they would in a pattern.
Syllogism	A form of reasoning in which two statements or premises are made and a logical conclusion is drawn from them.
Venn Diagram	Circle diagrams which show relationships between sets of items.

About the Authors...

Linda Sue Brisby	M.A. - Indiana University; B.A. - Goshen College; Reading Specialist - University of Michigan; 19 years of teaching experience in primary grade levels.
Andy Heidemann	B.A. - Ohio Wesleyan University; 6 years of teaching experience in upper elementary and junior high.
Natalie Hernandez	B.A. - University of Illinois; 22 years of teaching experience in various middle grade levels.
Jeanette Lenger	B.S. - Cal Poly State University, San Luis Obispo; 23 years of teaching experience in primary grade levels.
Ron Long	M.A. - Cal Poly State University, San Luis Obispo; B.S. - Long Beach State University; 21 years of teaching experience at the junior high level.
Petti Pfau	M.A. - Cal Poly State University, San Luis Obispo; B.A. - San Francisco State University; 17 years of teaching in middle elementary grades and in special education.
Scott Purdy	M.A. - Western State College, Colorado; B.A. - UCLA: 17 years of teaching/administration from middle elementary through high school level.
Sharon Rodgers	B.S. - Southern Oregon College; 26 years of teaching experience at all elementary through junior high levels.

Hands On Order Form

HANDS ON, INC.
2121 Rebild Drive
Solvang, CA 93463

NAME ______________________ DAYTIME PHONE ______________

ADDRESS ____

CITY ________________ STATE ____________ ZIP ________

NUMBER	BOOK TITLE	PRICE	TOTAL
______	Hands On Statistics, Probability, and Graphing	$15.95	________
______	Hands On Measurement	$15.95	________
______	Hands On Logic	$15.95	________
______	Hands On Geometry	$15.95	________
______	Hands On Algebra	$15.95	________
California residents please add 6% sales tax		TAX	________
Shipping: Please add $1.50 for the first book and $.50 for each additional book		SHIPPING	________

Please charge my VISA / Mastercard ________ / ________ / ________ / ________ ☐ Payment Enclosed ________

Expiration date ______ Cardholder's Signature ______________________________

Hands On Order Form

HANDS ON, INC.
2121 Rebild Drive
Solvang, CA 93463

NAME ______________________ DAYTIME PHONE ______________

ADDRESS ____

CITY ____________________ STATE ____________ ZIP ________

NUMBER	BOOK TITLE	PRICE	TOTAL
______	Hands On Statistics, Probability, and Graphing	$15.95	________
______	Hands On Measurement	$15.95	________
______	Hands On Logic	$15.95	________
______	Hands On Geometry	$15.95	________
______	Hands On Algebra	$15.95	________
California residents please add 6% sales tax		TAX	________
Shipping: Please add $1.50 for the first book and $.50 for each additional book		SHIPPING	________

Please charge my VISA / Mastercard ________ / ________ / ________ / ________

☐ Payment Enclosed ________

Expiration date ______ Cardholder's Signature ______________________________